THE

UN-UNIFIED FIELD

and

other problems

by MILES MATHIS

Introduction by
DR. T. YAQOOB

AuthorHouse™
1663 Liberty Drive
Bloomington, IN 47403
www.authorhouse.com
Phone: 1-800-839-8640

First published by AuthorHouse 6/28/2010

ISBN: 978-1-4520-0515-7 (e)
ISBN: 978-1-4520-0513-3 (sc)
ISBN: 978-1-4520-0514-0 (hc)

Library of Congress Control Number: 2010904528
Cover illustration by Miles Mathis
copy of Rubens' Daniel in the Lion's Den

Printed in the United States of America
Bloomington, Indiana

This book is printed on acid-free paper.

Dedicated to Diana
Queen of the Night

INTRODUCTION

Physics is now in crisis for a multitude of reasons. To start with, it is well known that the math blows up whenever quantum mechanics meets gravity. The failure to unify the two major maths of the 20th century has caused untold problems, and despite what we are sometimes told, the solution has not been found. Beyond that, particle physicists have convinced the world and large funding agencies that mass can be explained by a particle that itself has mass (the Higgs Boson). No one seems to see the *reductio ad absurdum* in that proposal. In a similar way, string theorists try to convince us that we live in eleven dimensions by telling us to take a hard look at a drinking straw.

In my own sub-field, astrophysicists have signed up wholesale to the belief that we only understand 96% of the mass-energy of the Universe. The consequent, stupendously large, one hundred and twenty orders of magnitude difference in the predicted and "observed" vacuum energy has been heralded by respected scientists as the most spectacular failure in the history of physics. Yet I have witnessed these very same scientists of repute laugh it off in conferences as some kind of whimsical rivalry between particle physicists and astrophysicists.

But all this is no joke: the problems are not going to be solved by wishful thinking, by fiat, or by fairy tales. Physics is in desperate need of new and fresh ideas, across the board. You will find plenty of these in Miles Mathis' book. I have not verified the results, but as far as I know, Miles is the first person to propose and begin to investigate the idea that Newton's fundamental gravitational equation already has electromagnetism embedded within it, and that what is needed is decomposition, not unification. Such a possibility has profound implications. To quote Michael Faraday, one of the greatest experimentalists in the history of physics, "If you would cause your view. . . to be acknowledged by scientific men, you would do a great service to science. If you would even get them to say 'yes' or 'no' to your conclusions it would help to clear the future progress. I believe some hesitate because they do not like their thoughts disturbed." I invite you to have your "thoughts disturbed" by Miles' book, if you dare.

Dr. T. Yaqoob astrophysicist at NASA and Johns Hopkins University*

*Disclaimer: this introduction in no way implies endorsement of any ideas in the book by either NASA or Johns Hopkins University.

CONTENTS

PREFACE

Are you tired of getting nonsensical answers from a physics that claims to know almost everything? If so, this book may be for you. Many years ago I noticed that no matter where I looked—on the internet, at the library, in books, on videos, in classrooms—I found little more than misdirection and fudged equations. I soon quit expecting anything from the normal sources and instead retired to the largest university libraries and top journals, hoping to find some real information there. But I found that the explanations in such places were even more convoluted, illogical, and incomplete than the explanations in the mainstream sources. I found no better answers, I only found denser and more heavily fortified answers. I found bigger maths and bigger claims of importance, but still found no clear and concise explanations of anything.

Primarily, this was because mechanics was completely missing. No one would ever tell me what a thing was or how it worked. No matter the topic, on the first page I was shunted immediately into differential equations or matrices or gauged fields or curved spaces. This math was never justified itself. I don't need to be taught math, I just need the author to give me a clear assignment of variables or spaces, but this was never done. Nor was the math ever tied to the mechanics. I have no fear of math. I like math. But math is a tool. In a physics paper, the math should be bookended by explanations. The first part of the paper should explain what the math hopes to achieve, and the last part of the paper should show what it has achieved. But the modern physics paper has jettisoned both the opening and closing, giving you only the free-floating math. The simple reason for this is that physicists no longer know what the hell they are doing. They can't explain anything sensibly, so they don't even try. They just unload a pile of math on you and expect that to keep you quiet for a while. If you complain, they browbeat you as someone who didn't take enough math in college. But the problem is not that the math is too hard, the problem is that the math is fake. If you study it, you find that it isn't showing what they claim to show. The math is nothing more than a false front, a parlor game, or a bad attempt at propaganda.

Unfortunately for them, some of us can see through even the foggiest equations. An important part of my research has been to unwind these equations, showing you the big fudges they contain. Some of you may have intuited that all this math was just camouflage, and I have proved that your intuition was correct. But it was even worse than any of you imagined. I know this because it was worse than I imagined, and I have a pretty vivid imagination. I assumed that these physicists, being human, were probably blowing smoke at least 50% of the time. After more than a decade of research, I can tell you that they are blowing smoke around 99% of the time. They claim to know almost everything, but they know almost nothing. The math is just a cover for this near-total nescience.

Unlike present theorists, I do not claim to be approaching a finished physics. I have no Theory of Everything, and my Unified Field, though a source of great pride, is also very limited. It has allowed me to climb a lot of walls in a short time, but I make no claims that it is a final theory or a perfect theory. Primarily, my work has allowed me to see through their claims, not to make big claims of my own. I see my central work as correcting their mistakes, and that itself is an infinite task. If I work on it everyday for the rest of my life, I will only make a small dent.

Contemporary physicists often claim that physics is almost over. Conversely, I believe that physics has just begun. It is a long drive into the unknown. But before you start a long drive, you have to back out of the driveway, and before you back out of the driveway, you have to clear all the snow and leaves and branches and toys and dogs and babies out from behind the car. Due to a century of neglect, we now have an awful mountain of debris behind our car, and the shoveling itself is the job of many years or decades. I am just a man with a shovel, taking care to separate the babies from the leaves and the snow.

UNIFIED FIELDS in DISGUISE

Both Newton's and Coulomb's famous equations are unified field equations in disguise. This was not understood until I pulled them apart, showing what the constant is in each equation and how it works mechanically.

A unified field equation does not need to unify all four of the presently postulated fields. To qualify for unification, it only has to unify two of them. The unified field equations that will be unmasked in this paper both unify the gravitational field with the electromagnetic field. This unification of gravity and E/M was the great project of Einstein and is now the great project of string theory. But neither Einstein nor string theory has presented a simple unified field equation. As time has passed this has seemed more and more difficult to achieve, and more and more difficult math has been brought in to attack the problem. But it turns out the answer was always out of reach because the question was wrong. We were seeking to unify fields when we should have been seeking to segregate them. We already had two unified field equations, which is why they couldn't be unified. We were trying to rejoin a couple that was already happily married.

Yes, both Newton and Coulomb discovered unified field equations. That is why their two equations look so much alike. But the two equations unify in different ways. Newton was unaware of the E/M field as we know it now, so he did not realize that his heuristic equation contained both fields. And Coulomb was working on electrostatics, and likewise did not realize that his equation included gravity. So the E/M field is hidden inside Newton's equation, and the gravitational field is hidden inside Coulomb's equation.

Let's look at Newton's equation first.

$$F = GMm/r^2$$

We have had this lovely unified field equation since 1687. But how can we get two fields when we only have mass involved? Well, we remember that Newton invented the modern idea of mass with this equation. That is to say, he pretty much invented that variable on his own. He let that variable stand for what we now call mass, but it turns out he compressed the equation a bit too much. He wanted the simplest equation possible, but in this form it is so simple it hides the mechanics of the field. It would have been better if Newton had written the equation like this:

$$F = G(DV)(dv)/r^2$$

He should have written each mass as a density and a volume. Mass is not a fundamental characteristic, like density or volume is. To know a mass, you have to know both a density and a volume. But to know a volume, you only need to know lengths. Likewise with density. Density, like volume, can be measured only with a yardstick. You will say that if density and volume can be measured with a yardstick, so can mass, since mass is defined by density and volume. True. But mass is a step more abstract, since it requires *both* measurements. Mass requires density and volume. But density and volume do not require mass.

Once we have density and volume in Newton's equation, we can assign density to one field and volume to the other. We let volume define the gravitational field and we let density define the E/M field. Both fields then fall off with the square of the radius, simply because each field is spherical. There is nothing mysterious about a spherical field diminishing by the inverse square law: just look at the equation for the surface area of the sphere:

$$S = 4\pi r^2$$

Double the radius, quadruple the surface area. Or, to say the same thing, double the radius, divide the field density by 4. If a field is caused by spherical emission, then it will diminish by the inverse square law. Quite simple.

The biggest pill to swallow is the necessary implication that gravity is now dependent only on radius. If gravity is a function of volume, and no longer of density, then gravity is not a function of mass. We have separated the variables and given density to the E/M field, so gravity is no longer a function of density. If gravity is a function of volume alone, then with a sphere gravity is a function of radius, and nothing else.

It is only the compound or unified field that is a function of mass. Yes, Newton's equation still works like it always did, and in that equation the total force field is a function of mass. But in my separated field, solo gravity is not a function of mass. It is a function of radius and radius alone.

Now we only need to assign density mechanically. I have given it to E/M, but what part of the E/M field does it apply to? Well, it must apply to the emission. Newton's equation is not telling us the density of the bodies in the field, it is telling us the density of the emitted field. Of course, one is a function of the other. If you have a denser Moon, it will emit a denser E/M field. But, as a matter of mechanics, the variable D applies to the density of the emitted field. It is the density of photons emitted by the matter creating the unified field.

Finally, what is G, in this analysis? G is the transform between the two fields. It is a sort of scaling constant. As we have seen, one field—solo gravity—is determined by the radius of a macro-object, like a Moon or planet or a marble. The other field is determined by the density of emitted photons. But these two fields are not operating on the same scale. To put both fields into the same equation, we must scale one field to the other. We are using both fields to find a unified force, so we must discover how force is transmitted in each field. In the E/M field, force is transmitted by the direct contact of the photons. That is, the force is felt at that level. It can be measured from any level of size, but it is being transmitted at the level of the photon. But since gravity is now a function of volume alone, it is not a function of photon size or energy. It is a function of matter itself; that is, of the atoms that make up matter. Therefore, G is a scaling constant between atoms and photons. To say it another way, G is taking the volume down to the level of size of the density, so that they may be multiplied together to find a force. Without that scaling constant, the volume would be way too large to combine directly to the density, and we would get the wrong force. By this analysis, we may assume that the photon involved in E/M transmission is about G times the proton, in size.

Now we continue on to Coulomb's equation:

$F = kq_1q_2/r^2$

One hundred years after Newton, we got another unified field equation. Here we have charges instead of masses, and the constant is different, but otherwise the equation looks the same as Newton's. Physicists have always wondered why the equations are so similar, but until now, no one really knew. No one understood that they are both the same equation, in a different disguise.

Coulomb's constant is another scaling constant, like G. Instead of scaling smaller, like G, k scales larger. Coulomb's constant takes us up from the Bohr radius to the radius of macro-objects like Coulomb's pith balls. It turns the single electron charge into a field charge.

But where is the gravitational field in Coulomb's equation? If we study charge, we find that it has the same fundamental dimensions as mass. The statcoulomb* has dimensions of $M^{1/2} L^{3/2} T^{-1}$. [Mass, Length, Time]. This gives the total charge of two particles the cgs dimension ML^3/T^2 . But mass has the dimensions L^3/T^2,** which makes the total charge M^2. So we can treat Coulomb's charges just like Newton's masses. We write the equation like this:

$F = k(DV)(dv)/r^2$

Once again, the volume is the gravitational field and the density is the E/M field. The single electron is in the emitted field of the nucleus, and D gives us the density of that field. But this time the expressed field is the E/M field and the hidden field is gravity. So we have to scale the electromagnetic field up to the unified field we are measuring with our instruments.

If k and G had been the same number, all this would have been seen earlier. It would have then been easy to see that Coulomb's equation was just the inverse of Newton's equation. But because the constants were not the same number, the problem was hidden.

In scaling up and scaling down, we don't simply reverse the scale. It is a bit more complex than that, as you have seen. In scaling down, we go from atomic size to photon size. In scaling up, we go from atomic size to our own size.

Unifying the two major fields of physics like this must have huge mathematical and theoretical consequences. Because Coulomb's equation is a unified field equation, gravity must have a much larger part to play in quantum mechanics and quantum electrodynamics. In a later chapter, I will show that gravity is 10^{22} times stronger at the quantum level than the standard model believes. Gravity must also move into the field of the strong force, and require a complete overhaul there.

By the same token, the E/M field must invade general relativity, requiring a complete reassessment of the compound forces. At all levels of size, we will find both fields at work, creating a compound field in which each field is in opposition to the other.

*This analysis also works with the Coulomb and SI.
**Maxwell showed this in article 5 [chapter 1] of his *Treatise on Electricity and Magnetism*. $a = m/r^2$, $s = at^2/2$, $m = 2sr^2/t^2$

A DISPROOF of NEWTON'S FUNDAMENTAL LEMMAE

Newton published his *Principia* in 1687. Except for Einstein's Relativity corrections, the bulk of the text has remained uncontested since then. It has been the backbone of trigonometry, calculus, and classical physics and, for the most part, still is. It is the fundamental text of kinematics, gravity, and many other subjects.

In this chapter I will show a simple and straightforward disproof of one of Newton's first and most fundamental lemmae, a lemma that remains to this day the groundwork for calculus and trigonometry. My correction is important—despite the age of the text I am critiquing—due simply to the continuing importance of that text in modern mathematics and science. My correction clarifies the foundation of the calculus, a foundation that is, to this day, of great interest to pure mathematicians. In the past half-century prominent mathematicians like Abraham Robinson have continued to work on the foundation of the calculus (see Non-standard Analysis). Even at this late a date in history, important mathematical and analytical corrections must remain of interest, and a finding such as is contained in this paper is crucial to our understanding of the mathematics we have inherited. Nor has this correction ever been addressed in the historical modification of the calculus, by Cauchy or anyone else. Redefining the calculus based on limit considerations does nothing to affect the geometric or trigonometric analysis I will offer.

The first lemma in question here is Lemma VI, from Book I, section I ("Of the Motion of Bodies"). In that lemma, Newton's provides the diagram below, where AB is the chord, AD is the tangent and ACB is the arc. He tells us that if we let B approach A, the angle BAD must ultimately vanish. In modern language, he is telling us that the angle goes to zero at the limit.

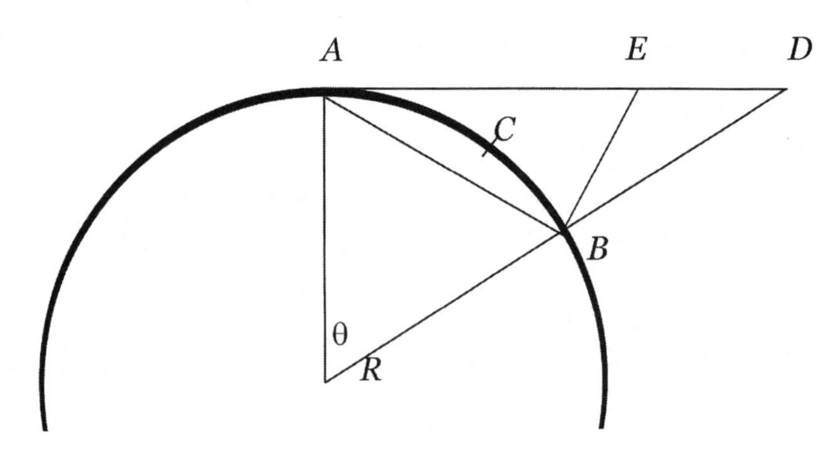

This is false for this reason: If we let B approach A, we must monitor the angle ABD, not the angle BAD. As B approaches A, the angle ABD approaches becoming a right angle. When B actually reaches A, the angle ABD will be a right angle. Therefore, the angle ABD can never be acute.

If we are taking B to A and may not go past A, then the angle ABD has a limit at 90°. Therefore, I claim we should see what happens to AD and AB at the limit of the angle ABD, instead of Newton's limit for angle BAD. In other words, I propose that Newton has monitored the wrong angle.

[If you are having trouble visualizing the manipulation here, it is very simple: you must slide the entire line RBD toward A, keeping it straight always. This was the visualization of Newton, and I have not changed it here.]

In Lemma VII, Newton's uses the previous lemma to show that at the limit the tangent, the arc and the chord are all equal. He proposes that because BAD goes to zero, AD and AB go to equality. I have just disproved this by showing that the angle ABD is 90° at the limit. If ABD is 90° at the limit, then ABD must always be greater than ADB. A triangle may not have two angles of 90°. Since AD is always longer than AB, the tangent must be greater than the chord, *even at the limit*. Please notice that if AB and AD are equal, then ABD must be less than 90°. But I have shown that ABD cannot be less than 90°.

1) If AB = AD, then the angle ABD must be less than 90°.
2) The angle ABD cannot be less than 90°.
QED: AB cannot equal AD.

In fact, this is precisely the reason that we can do calculations in Newton's "ultimate interval", or at the limit. If all the variables were either at zero or at equality, then we could not hope to calculate anything. Newton, very soon after proving these lemmae, used a versine equation at the ultimate interval, and he could not have done this if his variables had gone to zero or equality. Likewise, the calculus, no matter how derived or used, could not work at the limit if all the variables or functions were at zero or equality at the limit. In fact, the limit of ABD at 90° actually prevents the triangle or any of its parts from vanishing. The triangle gets smaller, but does not reach zero. Just as with the *epsilon/delta* proof, we get smaller but do not ever reach zero.

Some have tried to get clever and say that my claim that B never reaches A is like the paradoxes of Zeno. Am I claiming that Achilles never reaches the finish line? No, of course not. The diagram above is not equivalent to a simple diagram of motion. B is not moving toward A in the same way that Achilles approaches a finish line, and this has nothing to do with the curvature. It has to do with the implied time variable. If we diagram Achilles approaching a finish line, the time interval does not shrink as he nears the line. The time interval is constant. Plot Achilles' motion on an x/t graph and you will see what I mean. All the little boxes on the t-axis are the same width. Or go out on the track field with Achilles and time him as he approaches the finish line. Your clock continues to go forward and tick at the same rate whether you see him 100 yards from the line or 1 inch from line.

But given the diagram above and the postulate "let B go to A", it is understood that what we are doing is shrinking *both* the time interval and the arc distance. We are analyzing a shrinking interval, not calculating motion in space. "Let B go to A" does not mean "analyze the motion of point B as it travels along a curve to point A." It means, "let the arc length diminish." As the arc length diminishes, the variable t is also understood to diminish. Therefore, what I am saying when I say that B cannot reach A is that Δt cannot equal zero. You cannot logically analyze the interval all the way to zero, since you are analyzing motion and motion is defined by a non-zero interval.

The circle and the curve are both studies of motion. In this particular analysis, we are studying sub-intervals of motion. That subinterval, whether it is applied to space or time, cannot go to zero. Real space is non-zero space, and real time is non-zero time. We cannot study motion, velocity, force, action, or any other variable that is defined by x and t except by studying non-zero intervals. The ultimate interval is a non-zero interval, the infinitesimal is not zero, and the limit is not at zero.

The limit for any calculable variable is always greater than zero. By calculable I mean a true variable. For instance, the angle ABD is not a true variable in the problem above. It is a given. We don't calculate it, since it is axiomatically 90°. It will be 90° in all similar problems, with any circles we could be given seeking a velocity at the tangent. The vector AD, however, will vary with different sized circles, since the curvature of different circles is different. In this way, only the angle ABD can be understood to go all the way to a limit. The other variables do not. Since they yield different solutions for different similar problems (bigger or smaller circles) they cannot be assumed to be at a zero-like limit. If they had gone all the way to some limit, they could not vary. A function at a limit should be like a constant, since the limit should prevent any further variance. Therefore, if a variable or function continues to vary under a variety of similar circumstances, you can be sure that it is not at its own limit or at zero. It is only dependent on a variable that is.

If AB and AD have real values at the limit, then we should be able to calculate those values. If we can do this we will have put a number on the "infinitesimal." In fact, we do this all the time. Every time we find a number for a derivative, we put a real value on the infinitesimal. When we find an "instantaneous" velocity at any point on the circle, we have given a value to the infinitesimal.

Remember that the tangent at any point on the circle stands for the velocity at that point. According to the diagram above, and all diagrams like it, the tangent stands for the velocity. That line is understood to be a vector whose length is the numerical value of the tangential velocity. It is commonly drawn with some recognizable length to make the illustration readable, but if it is an instantaneous velocity, the real length of the vector must be very small. Very small but *not zero*, since we actually find a non-zero solution for the derivative. The derivative expresses the tangent, so if the derivative is non-zero, the tangent must also be non-zero.

Some have said that since we can find sizable numbers for the tangential velocity, that vector cannot be very small. If we find that the velocity at that point is 5 m/s, for example, then shouldn't the velocity vector have a length of 5? No, since by the way the diagram is drawn and defined, we are letting a length stand for a velocity. We are letting x stand for v. The t variable is not part of the diagram. It is implicit. It is ignored. If we are letting B approach A, then we are letting t get smaller. A velocity of 5 only means that the distance is 5 times larger than the time. If the time is tiny, the distance must be also.

Conclusion

My finding in this paper affects many things, both in pure mathematics and applied mathematics. I have proven, in a very direct fashion, that when applying the calculus to a curve, the variables or functions do not go to zero or to equality at the limit. This must have consequences both for General Relativity, which is tensor calculus applied to very small areas of curved space, and quantum electrodynamics, which applies the calculus in many ways, including quantum orbits and quantum coupling.

QED has met with problems precisely when it tries to take the variables down to zero, requiring renormalization. My analysis implies that the variables do not physically go to zero, so that the assumption of infinite regression is no more than a conceptual error. The mathematical limit for calculable variables—whether in quantum physics or classical physics—is never zero. Only one in a set of variables goes to zero or to a zero-like limit (such as the angle 90° in the problem above). The other variables are non-zero at the limit. For QM or QED, this means that neither time nor length variables may go to zero when used in momentum or energy equations.

This is not to say that length and time must be quantized; it is only to say that in situations where energy is found empirically to be quantized, the other variables should also be expected to hit a limit above zero. Quantized equations must yield quantized variables. Space and time may well be continuous, but our findings—our measurements or calculations—cannot be. Meaning, we can imagine shrinking ourselves down and using tiny measuring rods to mark off sub-areas of quanta. But we cannot calculate subareas of quanta when one of our main variables—Energy—hits a limit above these subareas, and when all our data hits this same limit. The only way we could access these subareas with the variables we have is if we found a smaller quantum.

As I said, there has also been confusion on this point in the tensor calculus. In section 8 of Einstein's paper on General Relativity[1], he gives volume to a set of coordinates that pick out a point or an event. He calls the volume of this point the "natural" volume, although he does not tell us what is "natural" about a point having volume. General Relativity starts [section 4] by postulating a point and time in space given by the coordinates dX_1, dX_2, dX_3, dX_4. This set of coordinates picks out an event, but it is still understood to be a point at an instant. This is clear since directly afterwards another set of functions is given of the form dx_1, dx_2, dx_3, dx_4. These, we are told, are the "definite differentials" between "two infinitely proximate point-events." The volume of these differentials is given in equation 18 as $d\tau = \int dx_1 dx_2 dx_3 dx_4$.

But we are also given the " natural" volume $d\tau_0$, which is the "volume dX_1, dX_2, dX_3, dX_4". This natural volume gives us the equation 18a: $d\tau_0 = \sqrt{-g}d\tau$

Then Einstein says, "If $\sqrt{-g}$ were to vanish at a point of the four-dimensional continuum, it would mean that at this point an infinitely small 'natural' volume would correspond to a finite volume in the co-ordinates. Let us assume this is never the case. Then g cannot change sign. . . . It always has a finite value."

According to my disproof above, all of this must be a misuse of the calculus, a misuse that is in no way made useful by importing tensors into the problem. In no kind of calculus can a set of functions that pick out an point-event be given a volume—natural, unnatural, or otherwise. If dX_1, dX_2, dX_3, dX_4 is a point-event in space, then it can have no volume, and equation 18a and everything that surrounds it is a ghost.

In the final analysis this is simply due to the definition of "event". An event must be defined by some motion. If there is no motion, there is no event. All motion requires an interval. Even a non-event like a quantum sitting perfectly still implies motion in the four-vector field, since time will be passing. The non-event will have a time interval. Every possible event and non-event, in motion and at rest, requires an interval. Being at rest requires a time interval and motion requires both time and distance intervals. Therefore the event is completely determined by intervals. Not coordinates, intervals. The point and instant are not events. They are only event boundaries, boundaries that are impossible to draw with absolute precision. The instant and point are the beginning and end of an interval, but they are abstractions and estimates, not physical entities or precise spatial coordinates.

Some will answer that I have just made an apology for Einstein, saving him from my own critique. After all, he gives a theoretical interval to the point. The function dX is in the form of a differential itself, which would give it a possible extension. He may call it a point, but he dresses it as a differential. True, but he does not allow it to act like a differential, as I just showed. He disallows it from corresponding to (part of) a finite volume, since this would ruin his math. He does not allow $\sqrt{-g}$ to vanish, which keeps the "natural" volume from invading curved space.

Newer versions of this same Riemann space have not solved this confusion, which is one of the main reasons why General Relativity still resists being incorporated into QED. Contemporary physics still believes in the point-event, the point as a physical entity (see the singularity) and the reality of the instant. All of these false notions go back to a misunderstanding of the calculus. Cauchy's "more rigorous" foundation of the calculus, using the limit, the function, and the derivative, should have cleared up this confusion, but it only buried it. The problem was assumed solved since it was put more thoroughly out of sight. But it was not solved. The calculus is routinely misused in fundamental ways to this day, even (I might say especially) in the highest fields and by the biggest names.

[1]*Annalen der Physik,* 35, 1911

ANGULAR VELOCITY and ANGULAR MOMENTUM

One of the greatest mistakes in the history of physics is the continuing use of the current angular velocity and momentum equations. These equations come directly from Newton and have never been corrected. They underlie all basic mechanics, of course, but they also underlie quantum physics. This error in the angular equations is one of the foundational errors of QM and QED, and it is one of the major causes of the need for renormalization. Meaning, the equations of QED are abnormal due in large part to basic mathematical errors like this. Because they have not been corrected, they must be pushed later with more bad math: abnormal equations must be made *normal.*

Any high school physics book will have a section on angular motion, and it will contain the equations I will correct here. So there is nothing esoteric or mysterious about this problem. It has been sitting right out in the open for centuries.

To begin with, we are given an angular velocity ω, which is a velocity expressed in radians by the equation

$\omega = 2\pi/t$

Then, we want an equation to go from linear velocity v to angular velocity ω. Since $v = 2\pi r/t$, the equation must be $v = r\omega$.

Seems very simple, but it is wrong. In the equation $v = 2\pi r/t$, the velocity is not a linear velocity. Linear velocity is linear, by the equation x/t. It is a straight-line vector. But $2\pi r/t$ curves; it is not linear. The value $2\pi r$ is the circumference of the circle, which is a curve. You cannot have a curve over a time, and then claim that the velocity is linear. The value $2\pi r/t$ is an orbital velocity, not a linear velocity.

I show elsewhere that you cannot express any kind of velocity with a curve over a time. A curve is an acceleration, by definition. An orbital velocity is not a velocity at all. It cannot be created by a single vector. It is an acceleration.

But we don't even need to get that far into the problem here. All we have to do is notice that when we go from $2\pi/t$ to $2\pi r/t$, we are not going from an angular velocity to a linear velocity. No, we are going from an angular velocity expressed in radians to an angular velocity expressed in meters. There is no linear element in that transform.

What does this mean for mechanics? It means you cannot assign $2\pi r/t$ to the tangential velocity. This is what all textbooks try to do. They draw the tangential velocity, and then tell us that $v_t = r\omega$. But that equation is quite simply false. The value $r\omega$ is the orbital velocity, and the orbital velocity is not equal to the tangential velocity.

I will be sent to the *Principia*, where Newton derives the equation $a = v^2/r$. There we find the velocity assigned to the arc.[1] True, but a page earlier, he assigned the straight line AB to the tangential velocity: "let the body by its innate force describe the right line AB".[2] A right line is a straight line, and if Newton's motion is circular, it is at a tangent to the circle. So Newton has assigned two different velocities: a tangential velocity and an orbital velocity. According to Newton's own equations, we are given a tangential velocity, and then we seek an orbital velocity. So the two cannot be the same. We are GIVEN the tangential velocity. If the tangential velocity is already the orbital velocity, then we don't need a derivation: we have nothing to seek! If you study Newton's derivation, you will see that the orbital velocity is always smaller than the tangential velocity. One number is smaller than the other. So they can't be the same.

The problem is that those who came after Newton notated them the same. He himself understood the difference between tangential velocity and orbital velocity, but he did not *express* this clearly with his variables. *The Principia* is notorious for its lack of numbers and variables. He did not create subscripts to differentiate the two, so history has conflated them. Physicists now think that v in the equation $v = 2\pi r/t$ is the tangential velocity. And they think that they are going from a linear expression to an angular expression when they go from v to ω. But they aren't.

This problem has nothing to do with calculus or going to a limit. Yes, we now use calculus to derive the orbital velocity and the centripetal acceleration equation from the tangential velocity. But Newton used a versine solution in the *Principia.* And going to a limit does not make the orbital velocity equal to the tangential velocity. They have different values in Newton's own equations, and different values in the modern calculus derivation. They must have different values, or the derivation would be circular. As I said before, if the tangential velocity is the orbital velocity, there is no need for a derivation. You already have the number you seek. They aren't the same over any interval, including an infinitesimal interval or the ultimate interval.

This false equation $v_t = r\omega$ then infects angular momentum, and this is where it has done the most damage in QED. We use it to derive a moment of inertia and an angular momentum, but both are compromised.

To start with, look again at the basic equations

$p = mv$

$L = rmv$

Where L is the angular momentum. This equation tells us we can multiply a linear momentum by a radius and achieve an angular momentum. Is that sensible? No. It implies a big problem of scaling, for example. If r is greater than 1, the effective angular velocity is greater than the effective linear velocity. If r is less than 1, the effective angular velocity is less than the effective linear velocity. How is that logical?

To gloss over this mathematical error, the history of physics has created a moment of inertia. It develops it this way. We compare linear and angular energy, with these equations:

$$K = ½\ mv^2 = ½\ m(r\omega)^2 = ½\ (mr^2)\omega^2 = ½\ I\omega^2$$

The variable "I" is the moment of inertia, and is called "rotational mass." It "plays the role of mass in the equation."

All of this is false, because $v_t = r\omega$ is false. That first substitution is not allowed. Everything after that substitution is compromised. Once again, the substitution is compromised because the v in $K = ½mv^2$ is linear. But if we allow the substitution, it is because we think $v = 2\pi r/t$. The v in K = ½

mv^2 *cannot* be $2\pi r/t$, because K is linear and $2\pi r/t$ is curved. You cannot put an orbital velocity into a linear kinetic energy equation. If you have an orbit and want to use the linear kinetic energy equation, you must use a tangential velocity.

The derivation of angular momentum does the same thing

$$L = I\omega = (mr^2)(v/r) = rmv$$

Same substitution of v for $r\omega$. Because $v = r\omega$ is false, $L = rmv$ is false. But this angular momentum equation is used all over the place. I have shown that Bohr uses it in the derivation of the Bohr radius. This compromises all his equations. Because Bohr's math is compromised, Schrödinger's is too. This simple error infects all of QED. It also infects General Relativity. It is one of the causes of the failure of unification. It is one of the root causes of the need for renormalization. It is a universal virus.

The correction for all this is fairly simple, although it required me to study the *Principia* very closely. We need a new equation to go from tangential or linear velocity to ω. Newton does not give us that equation, and no one else has supplied it since then. We can find it by following Newton to his ultimate interval, which is the same as going to the limit. We use the Pythagorean Theorem.

As $t \to 0$, $\omega^2 \to v^2 - \Delta v^2$

and, $v^2 + r^2 = (\Delta v + r)^2$

So, by substitution, $\omega^2 + \Delta v^2 + r^2 = \Delta v^2 + 2\Delta vr + r^2$

$\Delta v = \sqrt{v^2 + r^2}) - r = \omega^2/2r$

$\boldsymbol{\omega = \sqrt{} [2r \sqrt{v^2 + r^2}) - 2r^2]}$

$\boldsymbol{r = \sqrt{} [\omega^4/(4v^2 - 4\omega^2)]}$

Not as simple as the current equation, but much more logical. Instead of strange scaling, we get a logical progression. As r gets larger, the angular velocity approaches the tangential velocity. This is because with larger objects, the curve loses curvature, becoming more like the straight line. With smaller objects, the curvature increases, and the angular velocity may become a small fraction of the tangential velocity.

This means that the whole moment of inertia idea was just a fudge, used to make $v = r\omega$. Historically, mathematicians started with Newton's equations, mainly $v = 2\pi r/t$, which they wanted to keep. To keep it, they had to fudge these angular equations. In order to maintain the equation $v = r\omega$, the moment of inertia was created. But using my simple corrections, we see that the angular momentum is not $L = mvr = I\omega$. The angular momentum equation is just $L = m\omega$. We didn't need a moment of inertia, we just needed to correct the earlier equations of Newton, which were wrong.

[1] Newton, *Principia,* Section II, Prop. IV, Theorem IV, Cor. 1.

[2] Newton, *Principia*, Section II, Prop. I, Theorem I.

How to Solve GENERAL RELATIVITY problems without the TENSOR CALCULUS *in about 1/100th the time*

I have taught my students some wonderful mathematical shortcuts over the years, but this one is probably the most useful and the least known. In fact, I am not sure it is known at all, and so this paper. It is a mathematical manipulation only and has no necessary physical meaning, but it confirms the postulates of Einstein in a beautiful manner nonetheless.

The postulate that supplied me with this shortcut is the famous equivalence postulate, which states that gravity and acceleration are mathematically equivalent. In his lovely elevator car in space, Einstein showed that there is no difference between acceleration up and gravity down. After reading this, I decided to take Professor Einstein literally, and I reversed the vector by hand in several famous problems. This vector reversal had the mathematical effect of turning the field equations inside out. The objects in the field got larger with time, but the space around them reverted to a rectilinear or Euclidean field. In this field I could do very simple math, achieving the same numbers as Einstein. Then, when I was finished, I just turned the vector back around, to suit myself.

To show you how simple it is, I will solve both of General Relativity's most famous problems before you get to the end of this page—and you will understand the solutions, too.

The first problem is the precession of the perihelion of Mercury. Einstein told us this was caused by extra curvature from the Sun, so I will solve by looking at light move from the Sun to Mercury. Like Einstein, I assume that any influences from the Sun will move at the speed of light. And I assume that the curvature will reveal itself through the gravitational field. To get this revelation, I just let the Sun and Mercury accelerate at their own known rates, and watch the curvature.

Since the light moves from Sun to Mercury, the Sun's acceleration happens after the emission, and is not part of the math. We only need to look at Mercury's acceleration. Once we turn the vector around, Mercury will move in all directions, but we need to look at only one of them. We look in the direction of the orbital motion, which is tangent to the Sun. This one motion will show us the curvature, as you can see in the diagram.

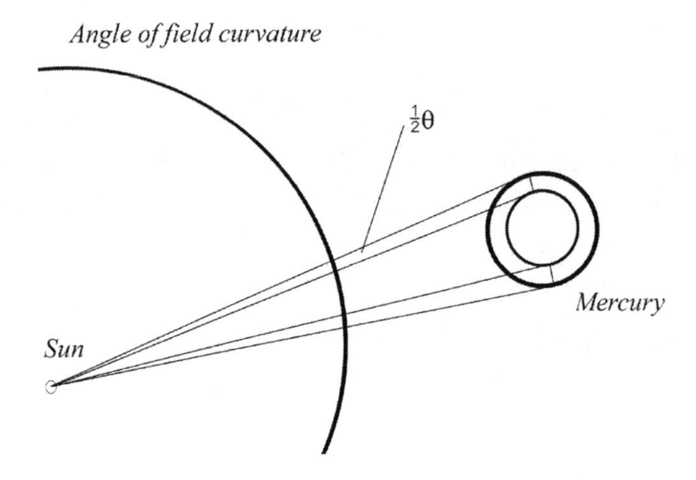

At perihelion, Mercury's orbital distance is about 46 million km. Light travels that distance in 153.3s. The acceleration at the surface of Mercury is 3.7m/s^2, so we find that one surface of Mercury will travel this far while the light is in transit:

$$s = (3.7\text{m/s}^2)(153.3\text{s})^2/2 = 43.5\text{km}$$

That gives us the field curvature in one direction, but we want the *total* field curvature. The orbit has extension both +x and –x in the field over each dt, so we must double our number. Then we just find the angle of curvature created by the total acceleration in that time.

$$\tan\theta = 87{,}000\text{m}/4.6\text{x}10^{10}\text{m}$$
$$\theta = .39 \text{ seconds of arc}$$

If we use the mean orbital distance instead of distance at perihelion, we get .49 arcsec. Einstein's field equations use both distances, since precession occurs at perihelion, but the curvature occurs over the entire orbit over a year. This is why his number, .45, just about splits the difference.

If you think that was a coincidence, watch the same solution applied to the bending of starlight by the Sun. We let the light pass the Sun and then travel to the Earth, where we see the bending. To calculate the amount of bending, we note the time it takes for the light to get here from the tangent, which is 501s. Once again, we reverse the acceleration vectors, but once again the Sun's acceleration happens after the light has passed. So we just look at the Earth's acceleration in 501s.

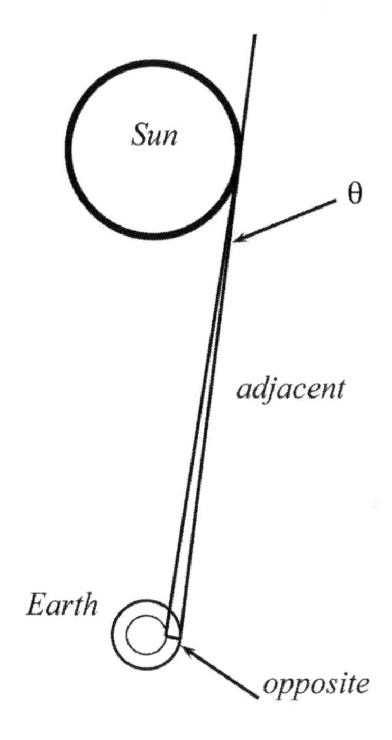

$s = (9.78 m/s^2)(501 s)^2/2 = 1{,}230{,}000 m$

$\tan\theta = 1{,}230{,}000 m / 1.51 \times 10^{11} m$

$\theta = 1.68$ seconds of arc

This is very near the number Einstein got (1.7), and this time we get the right amount without any doubling, since we don't need the entire field curvature here. Bending and precession aren't the same phenomenon. Mercury's light is getting bent in both orbital directions, since the planet as a whole is experiencing precession. But only the observer on the Earth is seeing the bending, so once we reverse the vector, it is his experience that is bent, not the light. As you can see from the diagram, the light is not curving. Only his measurement of it is. Since the observer is only on one position on the Earth, the bending uses only half the field. In order to see the curvature of the whole field, the observer would have to be on opposite sides of the Earth at the same time—an obvious impossibility.

All other General Relativity problems can be solved or estimated with this method, but I will let you discover that on your own.

ELECTRICAL CHARGE

Let us look freshly at the electrical field, as if we are encountering it for the first time—as if we are aliens just arriving upon Earth, uploading books from the Library of Congress and studying them for signs of intelligence. One of the first books[1] we open tells us this,

> The conceptual difficulties [of action at a distance] can be overcome with the idea of the field, developed by the British scientist Michael Faraday.... It must be emphasized, however, that a field is not a kind of matter. It is, rather, a concept*—and a very useful one.

Let us also look at the footnote referred to here:

> *Whether the electric field is "real" and really exists, is a philosophical, even metaphysical, question. In physics it is a very useful idea, in fact a great invention of the human mind.

As honest aliens, we must see that this footnote is absolutely false. The question of how the electrical field works is not a metaphysical one, it is a mechanical one. What Mr. Faraday has done is create a heuristic device that also works as a mental misdirection. Look at what the book says,

> The electric field at the location of the second charge is considered to interact directly with this charge to produce the force.

So the field was created in order to allow them to say that. Mr. Faraday, desiring to clear up action at a distance, drew a line with a pencil from one charge to the other, called that a field line, gave it no physical reality and no mechanical definition, and claimed that the pencil line interacted directly with the second charge. That is supposed to be a great invention of the human mind. As aliens, we must wonder what would be considered a dishonest creation of a human mind.

The book buries this bit of dishonesty with its own, calling mechanics "metaphysics." This is to warn off anyone from asking the questions we are asking here. Humans do not like to be called names, and these physicists are warning readers that if they ask any questions about the reality of the field, they will be called mean and scary names. Physicists do not like to be called philosophers or metaphysicians, and just the threat of it is enough to move them on. This despite the fact that the question involved is clearly and unambiguously one of mechanics. If transmitting a force from one object to another is not mechanics, nothing is.

Next we notice that the basic unit of electrical charge is either the Coulomb or, less often now, the statcoulomb. The Coulomb is defined as an Ampere-second, and an Ampere is defined as "that current flowing in each of two long parallel conductors 1m apart, which results in a force of exactly 2×10^{-7} N/m of length of conductor." We also are told that two point charges will feel a force of 9×10^9 N at a separation of 1m. A statcoulomb is "that charge on each of two objects that gives rise to a force of 1 dyne at the separation of 1cm."

$$1 \text{ statC} = 0.1 \text{ Am/c} \approx 3.33564 \times 10^{-10} \text{ C}$$

As aliens, our first question would be to ask what physical parameters the Ampere has, in terms of length and time. The books do not seem to want to tell us that. We are only told that the Ampere is used as a defining unit in order to obtain an "operational definition." This means that it is much easier to measure current than to measure charge, so the Earthlings have decided to base their units on current. But if we are persistent, we can discover electrical charge in terms of length and time. We have to go to the pages on the statcoulomb, and look at its historical derivation.

There we find Coulomb's law and the calculation of the constant in it.

$$F = kq_1q_2/r^2$$
$$k = 1/4\pi\,\varepsilon_0$$
$$\varepsilon_0 = 1/c^2\mu_0$$
$$\mu_0 = 4\pi \times 10^{-7} \text{ N/A}^2.$$

Wow, these Earthlings really don't want to tell us what they are doing. Here we have three different constants stacked on each other, each with a more obscure name than the last (electrostatic constant, permittivity of free space, permeability of vacuum) and the final constant circles back and is defined in terms of the Newton and the Ampere. We are being led on some sort of wild goose chase.

But we can still squeeze it out of them. Using Coulomb's law and working back from the last constant, we discover that *q* actually has no dimensions. All the necessary dimensions are given to the constants, and *q* is just a floater. If a Coulomb is an Ampere-second, then an Ampere must be a Coulomb per second. Go into the last constant equation, concerning the permeability, and substitute "nothing/second" for an Ampere. If you do that, you get the perfect dimensions for the force in Coulomb's equation. This means that electrical charge is mechanically undefined in the SI system. It cannot even be a wave, since a wave must be defined in cycles per second. We don't have any cycles here or seconds. An Ampere is a nothing per second, but a Coulomb is just a nothing.

The cgs system ditches the constant and gives the statcoulomb the dimensions $M^{1/2}\ L^{3/2}\ T^{-1}$. This gives the total charge of two particles the cgs dimension gm^3/s^2, which makes the total charge M^2. This gives the total charge of two particles the cgs dimension g^2 or m^6/s^4. **And this means that charge is mechanically and mathematically equivalent to mass.** Coulomb's equation is then not just similar to Newton's equation; it is exactly the same. We could actually write the charges as masses and nothing would change.

If we express one charge in terms of mass and one charge in terms of length and time, then Coulomb's equation gives us the force in gm/s^2.

$q_1 q_2/r^2 = F$
$(g)(m^3/s^2)/m^2 = gm/s^2$
But we could also express both charges is terms of mass
$(g)(g)/m^2 = F$
$F = g^2/m^2$

Or in terms of length and time
$F = m^4/s^4 = v^4$

Notice that this last equation tells us that a force is a velocity squared squared. That is perfectly logical, although it is not something we ever find in these physics textbooks.

Wikipedia, under the heading "statcoulomb", will not tell us that charge is the same as mass. Not only will it not admit that charge is dimensionally the same as mass, it goes out of its way to hide it. In this, Wiki is like all other standard textbooks, online and off. It says, "Performing dimensional analysis on Coulomb's law, the dimension of electrical charge in cgs must be $[\text{mass}]^{1/2}$ $[\text{length}]^{3/2}$ $[\text{time}]^{-1}$". But, as I just showed, that dimensional analysis stops short of completion by one very important step. Wikipedia asks the question, tells you the answer, but tells you the wrong answer on purpose.

Now, we must move ahead and ask why the Standard Model spends so much time larking about with permittivity constants and so forth. This is also misdirection. Many people will see those constants and think that free space or the vacuum actually have permeability or permittivity. But the truth is these constants are just folderol. Space and the vacuum only take on characteristics when you fail to give characteristics to your primary qualities. Charge, like mass or length, is a primary quality. It is a quality you assign directly to matter to explain the interactions you see or measure. If you create a quality and then fail to assign it any dimensions, then its dimensions will revert to the vacuum; but not otherwise.

But it is foolish to create a quality like charge and then refuse to let it have dimensions. Why would any scientist create a fundamental quality, refuse to define it mechanically, and then allow its parameters to be soaked up by the vacuum? A vacuum is supposed to have no parameters and no qualities, by definition. If we are going to give the vacuum qualities, we might as well flip our terminology and start calling the vacuum matter and matter the vacuum. Matter is supposed to be something and the vacuum is supposed to be nothing. But it is now the fashion for both the Standard Model and new theories to assign characteristics to the vacuum instead of to matter. This is nothing short of perverse.

As you can see, it is the old statcoulomb that has a degree of transparency. The Coulomb is defined in the most roundabout way, and then a bunch of meaningless constants are piled on top of it, to obscure it. Why? Why are physics textbooks such a mess? And why are they so much worse now than they were a hundred years ago? Why has the statcoulomb been replaced by the Coulomb? Why have the explanations become more obscure rather than less? Why would physics choose to replace the statcoulomb with the Coulomb, and hide the definition of charge beneath such embarrassing piles of absolute garbage?

Let me show you some more misdirection. Wouldn't it have been more logical to explain the electrical field in the same general terms as the gravitational field? In both cases we have a basic force between two particles. In both cases we create a field to help explain it. Why then vary the logic when expressing these two fields in scientific language? Why choose to express the gravitational field in terms of mass and acceleration, and the electrical field in terms of charge?

Given two large bodies, we see an apparent attraction and we assign the cause to mass. Given two very small bodies, we see a repulsion and we assign the cause to charge. Why not assign it to mass? Or, to put it another way, with large objects we immediately assign the cause of the attraction to the matter involved. The matter either acts directly or creates the field, therefore we call the causation "mass." Why not do the same thing with small particles? Why avoid mass and matter so persistently? Why create this nebulous thing called charge and never allow it to be explained mechanically?

With gravity, we assign the term directly to the force. Gravity creates the force or is the force. Mathematically, gravity is an acceleration caused by the force.

$$g = F/m = N/kg$$

But the electrical field is expressed without mentioning either mass or acceleration. Instead we have a characteristic called charge, which is either equivalent in dimension to mass (in the case of the statcoulomb) or which has no dimensions (in the case of the Coulomb). Let us skip the Coulomb as a mechanical non-entity and focus again on the statcoulomb. Remember that the statcoulomb is defined as a force at a distance. Well, gravity is also a force at a distance. Or, a statcoulomb is that thing that causes a force at a distance. The charge is not the force or the distance. It is the cause of the force, and the distance just gives us the magnitude.

Again, the same can be said for gravity. With gravity, mass is not the force or the distance, it is the cause of the force, and the distance just gives us the magnitude of the acceleration.

$$m = F/g = N/a = N/m/s^2 = Ns^2/m = (Ns/m)(s)$$

You may ask, why did I go on to express mass like that? Well, watch this. The Ampere is also defined as 2×10^{-7} N/m. A Coulomb is an Ampere-second. Therefore a Coulomb is

$$1C = 2 \times 10^{-7} Ns/m$$

So mass may be thought of a Coulomb-second.

The problem with all this is that using current definitions, a Coulomb has no dimensions or the dimensions of mass/second. But a statcoulomb has the dimensions of mass.

$$statC = L^3/T^2$$
$$C = L^3/T^3$$

Can both be right?

It is clear that we need to forget about current and finally define the charge mechanically. We must know what physical interactions are causing the forces, in order to clean up this mess.

To do this, the first thing we may notice is that when speaking of the gravitational field, a force does not have to include the distance at which it is felt. A Newton at a distance of 1 meter is the same as a Newton at a distance of 10m. A Newton is a Newton. Admitting this, why do we have as part of the definition of a Coulomb that it is a force at a certain distance?

The reason, of course, is that the electrical force is caused by a large number of sub-particles and (according to my theory) the gravitational force is not. If we assume that a static repulsion is caused by the bombardment by a huge number of tiny particles, then the total force is a summation of the individual forces of those particles. To obtain this summation, we must know a particle density.

And *that* is why we need to know a distance and a speed, in order to calculate a charge using the present theory. The distance gives us an x-separation between the two objects in repulsion, and since we assume the density is constant or near constant, the y and z density must be the same as the x-density. This gives us the size of the "field" that is creating the force. The speed gives us the density of the field at a given dt. In this way, the electrical field acts as a third particle moving from one object to the other, imparting the force by direct contact. But this third particle is much less dense than the two main objects. It acts like a discrete gaseous object, moving from one place to another at a given speed. This speed is of course c.

To calculate the charge, you need to know the mass and the distance. You are given the speed, c. This allows you to calculate the momentum. Notice that the distance is actually used to calculate the mass, since distance is telling you how large your gaseous object is. The distance is not telling you that you have a force working through a distance, as with the definition of the Joule. No, the distance is in the denominator in this case. You are dividing the force by the distance, and this is because you are seeking the mass of your gaseous object.

The speed, c, is used to calculate the mass of your gaseous object. Once again, this is because it is possible to calculate a mass if you are given a size (the distance) and a speed. The speed tells you the density at each dt. It is like a wave density. You have a certain number of sub-particles impacting your main particle at each interval. If you are given a length and a speed, then you have a time. This gives you a density.

You will say, yes, if you already know the force, then you can work back to find a mass for your gaseous object. You can find the mass of the electrical field that way. But if we don't know the force, then we can't know the mass, since we have no way of knowing the mass of each sub-particle. We must have something to sum, in order to find a density. If we don't know what each sub-particles weighs, we have nothing to sum. The speed and distance don't help us.

That is true as far as it goes, but the fact is that we *can* measure the force. That is why modern physicists have chosen to define everything in terms of the current. We can measure the force and the time and the distance. We know the speed also. Therefore it is quite easy to calculate the mass of the electrical field.

You will say, OK, but we still cannot know the mass of each sub-particle, since we don't know how many there are.

Once again true, but not really to the point. My point with this paper is not to assign a definite mass to the force-carrying sub-particle of the electrical field. It is to show that by giving mass to the electrical field we can totally dispense with charge, both the name and the idea. Charge is not a separate characteristic of matter. Charge is in fact the summed mass of these sub-particles.

[See chapter 22 for a simple way to calculate the mass of the charge photon.]

This allows us to clean up the great mess of the electrical field. Rather than define a fundamental characteristic like charge by later interactions, we can resolve that characteristic into even more fundamental characteristics. It is topsy-turvy to define charge in terms of current, since charge is supposed to be the cause and current the effect. You do not define causes in terms of effects. My housecleaning defines charge in terms of mass, which not only puts a floor under something that was hanging—it also allows us to throw the hanging thing out as garbage. It allows for a great simplification of theory.

Not only that, but it allows us to throw out a lot of meaningless constants at the same time. By assigning mass to matter in the field, we avoid having to assign characteristics to the vacuum or to free space. Free space does not have permeability or permittivity or anything else. Free space is free space. It is space, and it is free. It it were permeable or permittive, it would be neither. Only when you refuse to assign parameters to charge does free space begin to take on characteristics. Only when you refuse to make sense about matter, does your space also refuse to make sense.

Now we are in a position to resolve the Coulomb and the statcoulomb. Above I found that using only the dimensions of length and time

$\text{statC} = L^3/T^2 = M$
$C = L^3/T^3 = M/T$

Since I have shown how the mass of the radiation is calculated from the length and the speed, we can see where the difference comes from in these two equations. The statcoulomb comes directly out of Coulomb's equation. In that equation we are finding a single force. It has been called an instantaneous force, but since I don't believe in instantaneous forces, I will call it a force over one defined interval. Since it is force over one interval, we are dealing with a velocity, not an acceleration. You cannot have an acceleration over one interval. That is why the first equation has one less time dimension in the denominator.

But remember that we took the Coulomb equation from an experiment that measured current in a length of wire. Since we have an extended length, we must also have an extended time. Although we may have a constant velocity and therefore an acceleration of zero, we still must represent that series of intervals in our math. That is why the Coulomb equation has the extra time variable in the denominator.

Before I move on, let me clear up one other mess. The permittivity of free space is

$\varepsilon_0 = 1/c^2\mu_0 = 8.8541878176 \text{ x } 10^{-12}\ C^2/Jm$

Permittivity ε_0 is the ratio D/E in vacuum.

μ_0 is the permeability of vacuum, and has the value $4\pi \text{ x}10^{-7}N/A^2$.

N/A^2 turns out to be m^2/N, so that

$\varepsilon_0 = 8.85 \text{ x}10^{-12}kg/m^3$

Or, if we express mass in terms of length and time, then

$\varepsilon_0 = 8.85 \text{ x } 10^{-12} /t^2$

Why did I express the constant that way? One, to reduce it to its simplest dimensions. Two, to show that it can be assigned to something else entirely. Since free space cannot have permittivity, by definition of "space" and of "free," that constant must be owned by something else in the field. That number is not coming from nowhere, so some real particle or field of particles must own it. To discover what it is, we notice that it looks like an acceleration that lacks a distance in the numerator. We want to transform that number into an acceleration, so we need meters in the numerator and we need the denominator to be in seconds, rather than just "t^2". So we start by multiplying by 1 meter. That gives us a sort of acceleration, but we aren't allowed to just multiply by 1 meter without a transform. We must insert the meter into the equation in a legal manner, you see. To do that, we must ask how the time we already had in the equation and the meter we just inserted are related to each other. How many meters are in a second? Seems pretty difficult until we remember that light knows the answer. Light goes 300 million meters in a second, and that is the answer. In one meter, there are 1/c seconds, so we multiply by 1/c. That will allow us to insert the meter into the equation legally, and also give us seconds in the denominator. We end up with an acceleration of $2.95 \times 10^{-20} m/s^2$. That is lovely, because I have shown that is about the value of gravity for the proton. Yes, ε_0 is not the permittivity of free space, it is the gravity field created by protons (and electrons).

That clears up a lot of things, but let us look even more closely at the dimensions of the field in QED. The displacement field D is measured in units of C/m^2, while the electric field E is measured in Volts/m. As Wikipedia says, "D and E represent the same phenomenon, namely, the interaction between charged objects. D is related to the charge densities associated with this interaction, while E is related to the forces and potential differences."

$V = J/C = Nm//Ns/m = m^2/s$

If length and time are mathematically equivalent, as Minkowski taught, then we may reduce even further:

$V = J/C = Nm//Ns/m = m^2/s = m$

So a potential difference is just a distance, like any other difference.

$\varepsilon_0 = D/E = N/m^2///m^2//s/m = kg/sm^2 = m/s^3 = 1/s^2$

So you can see that, no matter how we juggle these equations and dimensions, we find that the constants are misleading. They tell us meaningless or contradictory things. But if we change two words in the sentences above from Wiki, we can get a clearer picture of the two fields D and E. Let us change the word "charge" to "mass."

"D and E represent the same phenomenon, namely, the interaction between massive objects. D is related to the mass densities associated with this interaction, while E is related to the forces and potential differences."

Now, if we were talking about a gravitational field or any other field, and you said that you divided a mass density (or just a density) by a force or potential difference, you wouldn't thereby create a permittivity or permeability in your vacuum or your free space. The simple act of creating or theorizing densities and forces does not create a resistance in the vacuum. The gravitational field has densities and forces and potential differences, and yet the gravitational field requires no resistance. Why? Simply because Newton was kind enough to assign dimensions to his characteristic "mass." He did not create a characteristic and then refuse to give it a dimension. Mechanically, his definition of mass is almost as empty as the definition of charge, but not quite. Newton tried to hide the fact that his mass was reducible to length and time by giving his constant a very strange dimension, but in the end these dimensions of G reduced to 1. This kept his constant just a fancy number, with no dimensions.

$G = L^3/MT^2$
$M = L^3/T^2$
$G = 1$

Since his constant has dimensions that are reducible to one, his field has no resistance or any other qualities. All the qualities are assignable directly to matter in the field.

The same is ultimately true of the electrical field, but physicists will not just come out and say so. In fact, they have preferred to imply, in the constants and fields they have created, that the vacuum does have characteristics. It has permittivity and permeability, which, if they really existed, would be types of resistance. But the electrical field has no resistance that it does not create itself, with the same matter that is creating the field in the first place. The only thing that resists the sub-particles are other sub-particles. The only thing that resists the gaseous object is other parts of the same object. The gas is material and it therefore resists itself. The radiation interferes with itself in a purely physical way, with no help from the vacuum.

When a Standard Model gas, made up of normal molecules, resists itself, we do not try to assign this resistance to the vacuum. We do not make up absurd abstractions like the permeability or the permittivity of the free space. We simply assign the resistance to molecule collisions. We could do the same thing with the electrical field, but we have so far preferred not to. Why?

Everyone knows that it is because once you admit that the E/M field is composed of radiation, you have to explain why the proton and electron aren't diminished by this radiation. We can create the sub-particle called the quark with no guilt or sin, since it doesn't immediately threaten to undermine the conservation of energy. But if the electrical field is composed of radiation, and if this radiation has mass, why doesn't the proton lose mass in radiating it? It is simply to avoid this question that the great mess of the electrical field has been left to sit. Physicists prefer a big mess and a big cover-up to an honest question.

My cosmology and mechanics answers this question in a very direct manner, without a lot of esoteric new theory. But despite the simplicity of the obvious answer, physicists are not interested in it since it requires they give up a lot of Standard Model gobbledygook that they have gotten very attached to. Piles of research money depend on sticking with the old assumptions, and money speaks louder than elegance or simplicity or logic.

Now let us show the first major outcome of my change in theory. If we give the radiation that causes the electric force the mass required to achieve this force, then we have a form of mass that must be opposed to the mass that creates the gravitational field. By that I mean that the two fields are in opposition to each other mechanically. One must be negative to the other. By this I do not mean anything esoteric. I am not creating some sort of mystical negative mass. I only mean to point out that every particle's radiation must have mass, and that this radiated mass creates a vector field that points out, whereas the gravitational mass points in. We already know that, in a sense. However, we have not included the idea in the math.

In other places I have shown that the E/M field is always repulsive, at the level of quanta. All forces are ultimately caused by bombardment. Electrical or magnetic attractions are always only apparent. This means that the proton does not actually attract the electron. It only repels it much less than it repels other protons. This leads to an apparent attraction, since the gravitational expansion of the proton allows it to capture the electron, but does not allow it to capture other protons. This leads to the appearance of attraction, in the dual field that is the gravity-E/M field.

When we measure the mass of a particle—either by using a scale or by looking at deflection—what we must be measuring is the sum of the two fields. We are measuring the gravitational force minus the force of the E/M radiation. This is simply because (to take the example of the scale) the radiation is bombarding our equipment, offsetting the "weight" of the particle itself. It is as if the particle is a little rocket, and our scale is the launchpad. The particle has it engines on all the time, and therefore we are not measuring the full weight of the particle. We are measuring the gravitational force minus the radiation force.

But this rocket analogy is not quite right, since a scale on the launchpad would actually measure the force of the exhaust. When we are calculating the mass of a particle, we are not putting it on a scale in that way. At the quantum level, we are measuring its deflections from other particles, and calculating its mass from the summed forces. But these forces must be compound forces. The expansion of the quantum particle makes it appear to attract all other particles; its radiation makes it repel all other particles. The total force is a vector addition of this attraction and repulsion.

What this means is that the true mass of the particle must be greater than the mass we measure or calculate with our instruments, whatever they are. If you take the mass of the particle to mean only its ponderable, gravitational characteristics, or only its force due to expansion, then that mass must be greater than the one we always measure. We are measuring the mass of the particle minus the mass of its radiation. Therefore its true mass is the measured mass *plus* the mass of the radiation.

Once you absorb that, it is time to consider the fact that calculating the true mass in this way must vastly increase the total mass of the universe. Over any dt, the mass of a given object is given by the expansion of the object in that time. But we can only measure the force due to expansion (gravity) minus the force due to the mass or momentum of all the radiation in that same time. Therefore the true mass must be the measured mass plus the mass of the radiation.

Also notice that this change in mechanics gives us a double addition of mass to the universe, since we gain both the mass of the radiation itself as well as the higher true mass of the radiating particle.

Both these statements are true:

1. The mass of the radiating particle must be greater than the mass measured by our instruments, since our instruments measure a compound mass.
2. The radiation itself has mass or mass equivalence due to energy, which is a second addition to the total mass of the universe. A radiating particle does not lose mass, which means that the "holes" left by radiation are filled by recycling the charge field.

Of course this immediately and simply explains the "mass deficit" in the universe and in current theory. We don't need massive amounts of dark matter or any other ad hoc fixes, since I have just shown the missing matter and energy. All we had to do is define our electrical field as a mechanical field instead of as pencil lines and we could have avoided this mess from the beginning.

[1] Douglas C. Giancoli, *General Physics*, p. 435

UNIFYING the PROTON and ELECTRON

This is another of the problems the standard model has failed to solve. QED and QCD do a lot of bragging, but they have very little to say about these fundamental questions. String theory also avoids simple questions like this, although these are the sort of questions a good quantum or atomic theory should answer first. We know that the electron weighs about 1820 times less than the nucleon, but after 90 years of experiment and theory, we still have no idea why. Once again, we have been told that the number 1820 is a fluke or a mystery, beyond physical comprehension, akin to the question of why horses have four legs instead of eight. They do, that is all. But as I will show, the number 1820 is not arbitrary or accidental. It can be arrived at by simple math and postulates.

My explanation begins by importing theory from chapter 8. There I showed that the mysteries of light motion and interaction could be explained by stacked spins, each spin outside the gyroscopic influence of inner spins. I showed the existence of four spins, of relative size 1, 2, 4, and 8, each orthogonal to neighboring spins. In other words, most photons are spinning every way they can spin, axially and in the x, y, and z planes. In my chapter on QCD, I will apply this to baryons, showing that baryons also have all possible spins. I will unify the proton and the neutron, showing that the difference between the two is only a difference in z-spin. That is, the particle at the center of every baryon is the same. Only the spins are different. In my chapter on mesons, I will show how this applies to them as well. Mesons are these same baryons stripped of outer spins. This unifies all hadrons. In this paper, I will show that the electron is also this same baryon stripped of outer spins. In this way, I will prove that electrons, mesons, neutrons and protons are all the same fundamental particle.

We begin this fundamental analysis by asking how the energy of a particle would increase when it goes from a state of no spin to a state of maximum spin. We start by arbitrarily assigning a non-spinning electron the energy 1. We also assign the number 1 to its radius. We do this because 1820 is a relative number, not an absolute number, so we don't care what the experimental values for mass are. We need only develop relative numbers. Obviously, the easiest way to do that is to start from a baseline of 1.

Next, we let the electron reach some small non-relativistic linear velocity v. That will be our baseline energy for the non-spinning state. To find how much energy the electron could gain by spin, we let the spin match the linear velocity. We let the tangential velocity of a point on the surface of the electron reach v. How much energy has the electron gained? Well, as the radius is to the velocity, the circumference will be to the spin. But we can't use $2\pi r$, since we must be looking at the tangential velocity, not the orbital velocity. The old equation $2\pi r/t$ applies (roughly) to the orbital velocity, but we can't use that since the energy of the electron or proton will be expressed mainly through its emitted field, and that field is emitted at a tangent, as a linear vector. We MUST use the tangential velocity here, which is why I have spent so much time in my theory separating the two mathematically and theoretically and developing a new equation for tangential velocity. What we find if we use my new equations is that the circumference is simply 8 times the radius. In kinematic or dynamic situations, we effectively replace π with 4. This gives us a spin energy of 8. We already had a non-spin energy of 1, so the total energy is 9. You may think of the non-spin energy as mass energy, or you may think of it as energy from linear velocity. Either way we must sum the two energies, since the total energy of the electron is a summation of spin and non-spin energies.

To clarify, we use the circumference here instead of the surface area, say, because we want the total energy of a given point on the surface of the electron. That point will have spin energy and non-spin energy. Given an axial spin of the electron, that point on the surface will have a vector at any given dt in one plane only. If we used the surface area equation, that would imply multiple vectors we don't yet have. We don't need to consider surface area until the next step, as you will see.

In this next step, we add the next spin, which is the x-spin. This spin is end-over-end, beyond the gyroscopic influence of the axial spin. Being end-over-end, this spin must have a radius or wavelength of 2. And since this spin is orthogonal to the axial spin, we now have too many vectors to use a simple circumference equation. We must switch to a sort of surface area equation. A point on the surface of our electron will now have a total of three linear vectors, one due to linear velocity, one due to axial spin, and one due to end-over-end motion. To express the total energy of the electron with x-spin, we use this term: [1 + (8 x 16)/2]. The radius is now 2, remember, so the 16 comes from 8r. The 8 comes from the axial spin, which we must multiply by the x-spin. We divide by 2 to express the fact that the particle itself is in the forward part of the x-spin only half the time, so only half the axial energy is affecting the x-energy in any one line of motion. What I mean is that the particle's x-spin will be moving against any linear motion half the time. A spin like this cannot combine with a linear vector by a straight addition. Only half of it can be expressed over any sum.

We repeat this same math and logic to create the y and z-spins. The radius of the y-spin is 4, so the term will be [1 + (8 x 16 x 32)/22]. We divide by 4 since we must use only half of both end-over-end spins. Likewise, the z-spin is [1 + (8 x 16 x 32 x 64)/24]. We divide by 2 squared squared because we are now in three dimensions. The x-spin is expressing only 1/4 of its strength relative to z, since it is orthogonal twice. The complete equation or representation then becomes:

[1 + 8], [1 + (8 x 16)/2], [1 + (8 x 16 x 32)/22], [1 + (8 x 16 x 32 x 64)/24]
= [1 + 8], [1 + 26], [1 + 211], [1 + 214] = 9, 65, 2,049, 16,385

The electron with all spins has an energy of 16,385. The electron with no spin has an energy of 1. The electron with axial spin has an energy of 9. If we divide 16,385 by 9 we get 16,385/9 = 1820.56

We may therefore deduce that the electron at rest is spinning only about its own axis. An electron with all possible stable spins is a proton, anti-proton, or neutron. An electron with no z-spin is a meson.

This number between the electron and the nucleon is very close to the atomic mass unit or Dalton, which has a value of 1822.

I will be asked how the electron can show a wave motion with only an axial spin. I have already shown that the wave characteristic of matter and of light is caused by stacked spins. But here we have only the first spin. How is the wave expressed? Well, it isn't expressed by an electron at rest, and we are comparing rest masses here. The electron must be moving to express a wave. If the electron begins moving and expresses a wave, of course it must have a second spin. It must get this spin from collision, we assume. And this second spin will add to the energy and therefore the apparent mass of the electron. A moving electron will become a sort of stable meson. As you can see from the math above, we can predict that it will have an energy about 7.2 times (65/9) that of the electron at rest. So in the first instance, the moving electron is not gaining energy only from Relativity. It is primarily gaining energy from x-spin.

UNIFYING the MESONS

Here I will show that mesons are protons or neutrons stripped of the z-spin, or electrons with superadded middle spins. I will show why mesons come in a variety of energies and how these energies are composed, using only stacked spins. I will also show why mesons are not stable.

In the previous chapter I found a basic meson state at 2049/9 = 227.67 times the energy of the electron at rest. This is an electron with 3 spins, or a proton missing the z-spin. This gives us 227.67 x .511 MeV = 116.3 MeV. That is not the energy of any known meson, so my first task is explaining how to arrive at a known meson from this number.

The Muon

Let us start with the muon, with a mass of 105.7 MeV. The muon is no longer considered a meson, but it is actually the primary and fundamental meson. It is the most stable meson, and it is the most stable *because* it is the most fundamental. It is simply the electron with three spins. It lacks only the z-spin. If it had a z-spin, it would be a baryon.

I have too much energy in my derived and predicted number of 116.3, and that is because that number includes the emission field. In chapter on electron/proton unification, I started at unity by assigning the non-spinning electron the number 1. I assigned no number to the emission field, so all my numbers in that analysis included it. In other words, my foundational theory postulated the particle and its emission as a single number, but experiment measures only the particle and its spins. In experiment, the emission escapes into the field (except with the neutron and other neutral mesons that trap the field). The muon has a charge and is not neutral, therefore its emission is not measured to be part of its mass or energy in experiment. This is the difference between my predicted number and the number of experiment.

Let us develop a number for that emission field. We remember from the spin equation in the previous chapter that the z level has 8 times the energy of the y level (16,385/2,049). But here we are looking at the emission field, not the spins of the particle. The two fields act in a reverse way. The total energy of the particle increases as we go to outer spins. The energy of the emission field increases as we move in. The emission field becomes denser as we get nearer the surface of the particle, while the spins get more energetic as they gain more radius and torque. But both fields follow the same radius and change in the same way, in reverse.

So we just need a mass for the charge field. It is simply the mass difference between the proton and neutron. The neutron traps this energy and the proton emits it. This energy is in the amount of 2.31×10^{-30} kg. But this energy is a field, so it must be denser if emitted from a smaller radius. If we multiply $8 \times 2.31 \times 10^{-30}$, we get 1.848×10^{-29}, which is 20.29 x the electron mass, which is 10.37 MeV. If we subtract that amount of emission from my predicted number, we get 116.3 - 10.37 = 105.9. That is near enough to the muon mass for my rough method here.

But why is there no neutral muon? By my mechanical theory of stacked spins, a neutral particle is a particle whose spins trap the emission. The neutron does this by creating a total spin path that sends the emission back to the particle (see illustration below). The emission cannot escape and becomes part of the mass of the particle. This is why the neutron weighs more than the proton. But the muon cannot create a path that comes back to the particle. We require four vectors for that, and the muon has only three. This is also why the muon has no stable state. The y-spin is the outer spin of the muon, and this spin must always be othogonal to the line of motion. The y-spin cannot match the linear motion, nor can it be opposite to it. So we cannot have a forward moving charge, creating stability, nor a reversed charge, creating neutrality. This missing neutral muon is convincing evidence in favor of my stacked spins. You can see that my theory has a simple mechanical explanation for it. Ask the standard model why there is no neutral muon.

The Pion

Now let us look at the pion, another common meson. The energy of the pion is about 139.6 MeV. In the previous chapter, I showed that the electron with axial spin only was the electron at rest (or low speed). To show a de Broglie wave motion, the electron had to move at a certain speed and experience collisions in the field. At this speed it would develop an x-spin and gain the energy of that spin. With both spins, the electron could express the wave in a simple mechanical way. I showed the energy of the electron with x-spin was 7.222 times the energy of the electron rest mass. I said that in this way the electron at speed was a sort of stable meson (since both x and y-spins could be called "mesonic" states). In the same way, I showed that the electron with y-spin would have 2,049 its rest energy. The electron with axial spin was the rest energy. If we divide 2,049 by 7.222 we obtain 283.7, which (multipying by .511MeV, the energy of the electron at rest) is 145 MeV. So, as a matter of stacked spins, the pion is the muon times 9 divided by 7.222.

Mechanically, we may imagine that the pion is created by a collision orthogonal to the direction of linear motion, stripping the baryon of z and x spins simultaneously, but momentarily boosting the y-spin. This boost is achieved by a linkage between the y-spin and the axial spin, as we see from the math. This is not difficult to postulate or visualize, since without the intervening x-spin, the a and y vectors would naturally link. Nor is it difficult to imagine a collision that would target x and z levels, since they are also linked as vectors.

This would also explain why a pion quickly becomes a muon, and why the muon is more stable. The muon has all inner spins, at natural levels. It only lacks the z-spin. It has a normal level of charge protection, although this charge is moving orthogonally to the linear motion, and therefore cannot express its full character. The charge field also lacks a z-spin, stripping it of its largest defense. Therefore the muon is not stable. But the pion is even less stable, since it is missing the x-spin as well as the z-spin. It is momentarily more massive, since it has gained energy from the collision. But this mass cannot be maintained, since the fields that could absorb it are gone. The pion tries to rebuild as a muon, funneling energy down from y to x, but it cannot do so. As the pion collapses, it pauses momentarily at the muon energy while the rebuilding is attempted and failed, then, like the muon, it dissolves. It, like the muon, must dissolve either into an electron with axial spin, which we can detect; or into an electron with no spin, which we cannot detect. We will not detect the failed particle until it benefits from collision and re-establishes itself as a spin particle (or forms a multiple and becomes a neutrino—see below).

Now, we only need explain the difference between 145 and 139.6. Fortunately, this gap is the same relative size as my gap with the muon. We had a 10.37 MeV gap with the muon, and we have a 5.4 MeV gap with the pion. That is near enough to 2 for this paper. The pion is larger, so by my math and theory, we would expect a smaller gap. The gap is caused by the emission field, so how would we expect the emission of the pion to differ from that of the muon? The pion is missing the x-spin on the emission. This means that it is missing one of the 2's in the math. Even though the y and a-spins are augmented by the collision, the y-spin has no x-spin to double beneath it (see the full spin math in my previous chapter for clarification here).

The biggest problem with the pion is found in the neutral pion. Experiment gives us an energy of about 135 for the neutral pion, and that cannot fit my theory at all, at first glance. I have proposed that neutral particles block emission, so that the mass of the emission must increase the mass of the particle. A neutral pion should act like a neutron, swallowing its own emission. This would make it weigh *more*. The neutron weighs more than the proton, not less. So the neutral pion should weigh more than the charged pions. A neutral pion should weigh 145, not 135. But this also has a simple mechanical explanation. Blockage of emission can only take place if the emission travels through the four spins and obtains a final trajectory that takes it back toward the particle. This is what happens with the neutron, as I will show in my analysis of beta "decay". But the pion cannot create this path. It lacks x and z spins. It has only a and y spins. Even though the a and y spins are augmented, they still cannot create the path in the right way. We have mechanical vectors at all points, with nothing esoteric or hidden, so if we want to explain a different energy, we must do so directly, with a clear visualization.

The answer lies in the easily demonstrable fact that a particle with only a and y spins will funnel the emission back to the particle, but the emission will miss the particle (see illustration below). Therefore, the emission is in vector opposition to the linear motion of the particle, but it is not "re-absorbed" by the particle. If the particle is moving in the +a direction, the emission will be summing in the -a direction; but since it is missing the particle, its energy cannot be added to the total. Its energy must be subtracted from the total. This is why we subtract the 5 MeV instead of adding it. Instead of 140 + 5, we obtain 140 - 5, which gives us about 135.

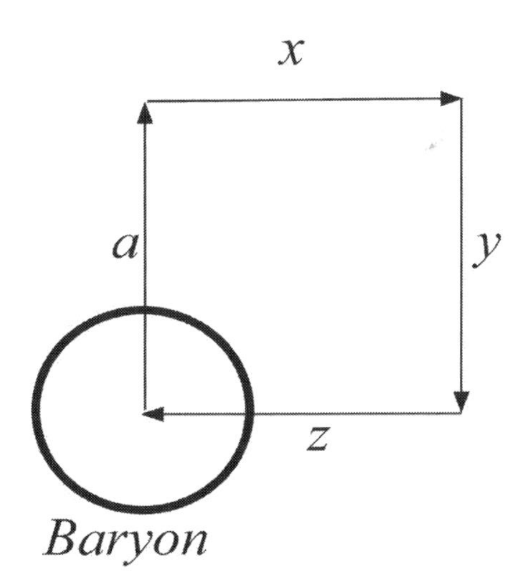

with only a and y vectors, the pion's emission misses the particle (the particle is also moving +a)

[In imagining the loss of x and z vectors in the illustration above, you must not allow the a and y vectors to line up, so that y points right back at the particle. This is because this two-dimensional illustration is missing the third dimension. In a three-dimensional schematic, the a and y vectors would not in fact line up. The illustration here is a useful visualization in some ways, but it is not complete or completely accurate.]

The Muon Neutrino

I proposed above that muons and pions must dissolve either into electrons with axial spin or without. We know what an electron with axial spin looks like, since I have proposed that this is the normal electron at rest. But what is an electron with no spin? In the chapter on electron/proton unification, I showed that the non-spinning electron would have a mass of 1/9 that of the spinning electron. But do we ever find such a particle? We wouldn't expect to find it by normal methods, since the normal methods use E/M fields, and the non-spinning electron would have neither charge nor magnetism. But watch this: If we divide the current energy of the electron .51MeV by 9, we obtain .05678., which is 56.78 keV. Still not ringing a bell? Well, the energy of the muon neutrino is 170keV. If we divide 170 by 56.78, we get exactly 3. There is no muon neutrino. The muon neutrino is three non-spinning electrons huddling together for protection, each one trying to block the charge wind in one of the 3 dimensions. These electrons have no charge or emission, so they do not repel each other. They cannot repel any other particles either, so they get pushed into groups like this, where their only repulsion is from the strength of their shells or surfaces.

All neutrinos may be explained by other means. Here I have shown that the muon neutrino is three non-spinning electrons in a group. In my QCD paper, I showed that the electron neutrino was a variation in the charge field caused by a z-spin reversal. The tau neutrino will be lost below in the same way, to a simple mechanical explanation.

Neutrino "Oscillation"

I showed at the end of my QCD paper that neutrino oscillation is actually a change in the B-photon or emission field, not the reincarnation of one particle to another. But here I give the muon neutrino to an electron triplet. Which is it? Both. In neutrino oscillation we go from one unseen flux to another. We call the first flux an electron neutrino and the second a muon neutrino. But in reality, the first flux is a flux of the B-photon field, with no neutrino present. The second is a flux of the electron field, with electrons colliding with free neutrons and losing their axial spins. Again, there is no neutrino present, but there is a new particle present after the second event: an electron triplet. In each stripping, a neutron is stripped of its z-spin and multiple electrons are stripped of their axial spins. It would appear that it takes three electrons to strip each neutron. The axial spin has a much lower energy than the z-spin, and only by adding their linear energy to the equation can the moving electrons successfully attack a neutron. I will have more to say about this in other papers.

The Tau Neutrino

The tau neutrino has been called a neutrino simply because the standard model had no other way of categorizing such a small meson. The tau neutrino has an energy of 15.5 MeV. That is about 30 times the energy of the resting electron. If we bring an electron up to speed, it will gain an x-spin and thereby a wave. I have shown that this x-spin will give the electron 7.222 times the energy of the resting electron, which is 3.69 MeV. You can see from both these comparisons that the tau is about 4 times the x-spinning electron. So it turns out that the tau is another multiple: it is four x-spinning electrons. But why would x-spinning electrons join in groups of four?

As it turns out, the x-spinning electron has a number of states, and only one of these states is stable. That is the electron we measure when we measure an electron with a wave. In this state, the emission is summing in the same direction as the linear motion. To say it another way, the linear vector and the x-vector match. This gives the moving electron a protective charge field in front with its full numerical value, and allows the electron to repel other electrons.

But this is only one of several possible states. In all analyses up to now, I have assumed that the electron had only four total vectors, and I have implicitly let the linear vector match the axial vector. But now I must include the further complication that the axial and linear vectors may be orthogonal. If this is the case, we will have to analyze not four but five orthogonal vectors in every baryon. For instance, let the electron spin about a y-axis. The emission will be thrown out mainly in the xz plane. If we let the electron now move in x or z, then the linear motion and the emission will match, in a vector sense. But if we let the electron move linearly in y, it will be moving orthogonally to its own emission, you see. It would not be stable, because it would have no protection from the emission. This unstable x-electron is the component of the tau neutrino.

To say it another way, the particle is moving forward and emitting sideways. The x-spin is unstable, and would tend to decay back to the axial level or lower. This is how we get the tau neutrino. The tau neutrino is four x-spinning electrons, all of the unstable variety, and all different from the others. By setting themselves up this way, the unstable x-spinning electrons are able to to become stable. The tau neutrino is emitting up, down, and to both sides, and has a charge weakness only directly ahead of it, where there is a minima. For this reason, it is stable, but not as stable as an electron. I have called all these constituents of the tau neutrino electrons, but I think it is clear that two of them are positrons. To be precise, I should say that the tau neutrino is made up of two electron/positron pairs, orthogonal to one another.

I should also come up with a better name for the tau neutrino, since I have now shown that it is not a neutrino. There are no neutrinos. I will call it the quirino for now. In honor of Majorana, who came so close to solving the problem of gravity. It is close to quarto, but I don't want quarto since it is already overused.

The Kaon

In experiment we find that a kaon in decay can yield either pions or muons. This was the hint I needed to discover its construction. The most stable kaon is very long-lasting, living for some 10^{-8} seconds, and weighing about 493.7 MeV. Two charged pions and two muons give us almost 492 MeV. The long life is given it by the muons. They add stability to the kaon, making it much more stable than other mesons composed only of pions (such as the eta meson, below, which lives only 10^{-19} s). Pions and muons both emit sideways, which allows them to huddle in groups of four, either with one another or with with themselves. We have four charge orientations at the y-level, and one particle in the quadruplet represents each orientation. The kaon therefore emits to all four sides, missing only the forward emission (relative to its linear motion). The lack of forward emission makes the kaon a meson, and keeps it from having stability. But the partial muon construction protects it in collision, since the muon has all inner spins.

The Eta Meson

The eta meson is clearly just 4 pions. It is either 3 neutral pions and a charged pion, or 2 neutral pions and 2 charged pions. This would give us either about 545 MeV or 550 MeV, depending on the combination. The eta meson is 547.8 MeV in experiment. Currently the eta is said to decay into three pions, but the trackers are missing one of the pions, due to the orthogonal vectors. It is almost impossible to track orthogonal vectors simultaneously in such a complex decay.

Charmed Eta Meson

The charmed eta meson is another meson that matches my equation outputs almost precisely. You will have noticed that my fundamental meson states are all multiples of 2. The charmed eta resides near this level: [1 + (8 x16 x 32 x 64 x 128)/29] = 65,536. We divide by 9 to achieve a multiple of the electron, and then multiply by .511, which gives us 3721 MeV. As with the pion, we only need one more factor, and that is 7.2222/9 (see above). This gives us 2986 MeV. The charmed eta meson in experiment is 2980. From this we may conclude that the charmed eta is related to the pion, each of them losing inner spins in collision.

Bottom Eta Meson

The bottom eta is another meson created like the pion and charmed eta. We have a fundamental meson level at [1 + (8 x16 x 32 x 64 x 128)/28] = 131,072. If we divide by 9, we obtain 14,560, and if we multiply by .511, we get 7,442 MeV. If we multiply by 9/7.222, we obtain 9273 MeV. The experimental value of the bottom eta is 9300 ± 40. Like the charmed eta and the pion, the bottom eta is missing inner spins after a collision from the side.

The Tau

The tau has been a very difficult particle to isolate in experiment, and that is because it is a complex combination of mesonic states. It decays into at least one muon, and this fact helped me a bit to discover its make-up. As I showed above, pions commonly decay into muons, as they shed unstable outer spins. So we should look for the pion in the tau. The tau was also a difficult particle for me to explain, since it is the first example I uncovered of a baryon huddling with a meson. In this case, we have two unstable baryons huddling with two pions.

Let me first explain the unstable baryon. In my QCD paper I offered 16 baryon states, assigning eight to neutrons/anti-neutrons and eight to protons/anti-protons. This was a complete list of stable baryons, but it is not a complete list of baryons. We have 16 more unstable baryons. The mechanical reason for this goes back to my analysis of the tau neutrino, where I show that the emission field of any emitting particle may be emitted sideways. I showed that if we include the linear motion as a fifth vector, this vector may be orthogonal to the outermost spin. Since the spin does not emit in the same direction as the linear motion, it does not provide full charge protection. So the baryon may be emitting to any of four sides, all of them orthogonal to the linear motion. If we combine these z-spins with inner spins, we obtain 16 more states. All of these 16 are unstable. This gives us a total of 32 baryon states, and that number can be arrived at easily by summing the five possible vectors. With five vectors, we must have 2^5 possible states.

The tau makes use of two of these unstable baryon states. You may imagine that these two opposite baryons meet back to back, trapping two pions in the other corners. In other words, if the baryons are emitting north and south, the pions emit east and west. This gives the tau a split second of stability, until the baryons begin to turn and face each other. The intervening pions prevent this turn for an instant, but they are overwhelmed by the energy of the baryons, and the structure quickly collapses. The pions become muons, and the baryons strip each other of spin completely, decaying back to spinless electrons, and becoming invisible to our detectors.

But there is one other complication. The baryons are doubly unstable, due to a recent collision. Just as pions are damaged particles, lacking inner x-spins, the baryons in the tau are damaged baryons. They have been hit from the side, losing this same x-spin. It is as if the particle has been crushed in a vise, by a field orthogonal to its main motion. This orthogonal field targets inner spins. So we multiply by that same term we have seen before with the pion: 7.222/9. This term express a loss of the x-spin, and a linkage of the a and y spins. So take the baseline energy for the baryon [1 + (8 x16 x 32 x 64)/24] = 16,384, divide by 9 to find a multiple of the electron and multiply by .511.

This gives us 930 MeV. Then we multiply by 7.222/9, to express the collision or vise. This gives us 746.6 MeV. Doubling that to express two particles gives us 1493. Two charged pions gives us 279.2. Adding them all together, we have 1772. The energy in experiment of the tau is 1777.

To sum up, we have two baryons and two pions, both damaged from the same sideways field or collision, and all missing the x-spin. They combine in a NSEW square, and then quickly decay.

The Z particle

Using my meson equation, we can find a value for the Z. The Z is a baryon with several unstable spins forced on top of the stable z-spin. Here is the primary equation: $[1 + (8 \times 16 \times 32 \times 64 \times 128 \times 256)/2^{14}] = 524{,}288$. We divide by 9 to achieve a multiple of the electron, and then multiply by .511, which gives us 29.8 GeV. Tripling this gives us 89.3 GeV. The Z in experiment is about 91.2 GeV. The difference is 1.9 GeV, which is almost exactly 18 muons. We know that we get muons from Z decay, we just aren't catching all of them. A Z decay should provide a veritable shower of muons, but the detections are nearly planar, so only a few of the tracks are seen at any one time. We see from this math that the Z is actually three large particles, not one. Like the muon neutrino, the Z is attempting to block the charge wind by huddling. Each of the three large particles is accompanied by 6 muons, so we have a sort of very unstable complex molecule here, one that it should be fun to model. The reason we have six can be seen from the equation. We have 2 x 3 muons for each of three large particles, and we have two levels above the stable baryon level in the math. The 64 level is the baryon level, and we have both spin levels above that of 128 and 256 here. That is the 2 in the 2 x 3. The 3 follows the 3 in 3 large particles, although you probably need to be looking at the model to see this clearly. The three large particles huddle, Therefore it should be expected to decay into three electrons, or the theoretical equivalent. But since the electrons will flee in three different planes upon breakup, they would be hard to detect simultaneously.

The D meson

"Meson" is not a logical term for a particle above the baryon mass, so I will call these particles "uberons." The D meson is another fundamental level predicted by my math. We use this equation: $[1 + (8 \times 16 \times 32 \times 64 \times 128)/2^{10}] = 32{,}768$. Dividing by 9 and multiplying by .511 gives us 1860.5 MeV. The D meson in experiment is about 1865. The math shows us that the D meson is a baryon with an added spin above the z-spin. We can add either a second x or a second y level above the z level, with one giving us 1860.5 MeV and the other giving us 7442 MeV (see the bottom eta). At 7442 MeV, the numerator is the same, but the denominator will be 2^8. The number of 2's depends on whether the outer spin is "doubling" the inner x or inner y level. One will be orthogonal twice and the other will be orthogonal three times, so the denominator must represent this inner doubling.

Other mesons and uberons

Several other mesons and uberons can be explained as multiples of the ones above. Using the tools I have created in this paper, we can find the structure of any meson or uberon. For instance, the vector B meson is just three taus. We can see this immediately from the energies. The rho meson is one baryon without x-spin and two tau neutrinos. The strange D meson is four kaons.

The charmed B meson is the first uberon state (7442 MeV) without inner x-spin, plus two pions. That gives us 5973 + 140 + 140 = 6253. The charmed B in experiment is 6275. Once again we have the baryon and pions missing the same x-spin, from the same sideways field.

The so-called vector mesons have much shorter lives than most of the others, and this is because they are often initial states of pseudoscalar mesons. The vector meson decays into the pseudoscalar meson on the way down. This does not mean that the vector mesons are taking part in quark bonding or anything else. They are not bosons. They are simply more complex multiples of the basic states shown above, which makes them more unstable and often larger.

Conclusion

Perhaps after all this you can see that the ad hoc quantum numbers of QCD are no longer needed. CP parity, for instance, is completely explained by stacked spins. We don't have "vector" mesons or "pseudoscalar" mesons, we simply have mesons and uberons. Inner spins give us a simple mechanical and visualizable explanation of parity, as well as all the other quantum numbers like spin, isospin, and flavor. My theory will do a housecleaning not only on the particles, but on the names. We can now jettison most of the useless and un-mechanical vocabulary of atomic physics. In upcoming papers I will suggest simplified terminology to go with our new simplified physics.

SUPERPOSITION

In this paper I will offer a simple mechanical explanation of superposition. I will provide an easy visualization as well, one that simultaneously solves the mystery of superposition and the wave motion of particles.

Heisenberg and Bohr assured everyone that this was not possible. The Copenhagen interpretation, which is still the preferred interpretation of quantum mechanics by contemporary physicists, states in no uncertain terms that the mysteries of quantum physics are categorically unsolvable. That is, they are not only unsolved, they are impossible to solve. All other interpretations of quantum mechanics have agreed with this interpretation, regarding the impossibility of a straightforward visualization or of a simple mechanical solution. Some variations have denied other aspects of the Copenhagen interpretation, especially regarding its opinion of the collapse of the wave function. Bohm, for instance, has attempted a deterministic explanation of certain parts of QED, including a reinterpretation of the wave function and of the Uncertainty Principle. But not even Bohm or Bell believed that anyone could offer a simple visualization that would explain superposition or the so-called wave-particle duality.

Einstein came closest to this belief. He remained convinced that quantum mechanics would eventually be explained in a more consistent manner. But, again, it was mainly the probabilistic nature of quantum dynamics that bothered him, not the fact that it could not yield to simple visualizations. He did not like God playing dice, but he did not expect God to draw us a picture with every new theory.

I did not approach the problem intending to find a visualization or an easy mechanical solution. I only wanted to make better sense of it in my own mind. But in analyzing the problem I found that the mechanical difficulties were not nearly as formidable as has been claimed. I found that I could quite easily visualize the physical motions, and that I could put these visualizations into pretty simple words and pictures. One basic discovery allowed me to do this, and that is what this paper is about.

I believe that the most efficient way to lead the reader through the problem is to analyze the current explanation of superposition, as it is presented in a contemporary text. As my text I will use David Albert's *Quantum Mechanics and Experience*. I choose this book for the same reason that the status quo chose to publish it: it puts the theory in as clear a form as possible, for laymen and physicists alike. Albert is a philosophy professor at Columbia, but he has been embraced and tutored by many mainstream physicists. This book may therefore be taken as a representative, if not perfect, expression of current theory. If it were not it surely would not have been published by Harvard University Press.

Albert begins by taking two measurable qualities of an electron. He tells us that the qualities don't matter, and that we could call them color and hardness if we wanted to. In a footnote on page 1 he informs the reader that experimentally he is talking about x-spin and y-spin, but he does not elaborate beyond that. Conveniently, this footnote allows me to make my first major substantive point. From a logical point of view, an electron cannot have angular momentum on the x and y axis at the same time—not if both spins are about an axis through the center (Albert claims that they are). Imagine the Earth spinning about its axis. Call that axis the x-axis. Now go to the y-axis, which also goes through the center but is at a 90° angle from the x-axis. Try to imagine spinning the Earth around that axis at the same time that it is spinning around the x-axis. If you can imagine it, then you have a very vivid imagination, to say the least. If that didn't convince you, then remember the gyroscope and the phenomenon called precession. A torque applied to the axis of rotation is deflected, so that circular motion is not allowed about the y-axis. You can have circular motion in only one of the two planes at a time. To see why this is so, think of a point on the surface of the sphere or on the edge of a wheel. Give it spin in the xy-plane. Now follow its course and see the curve it describes. Once you have done that, think of giving it a spin in the zy-plane at the same time. You have a second curve applied to the first curve. But these two curves cannot be added to create a new curve that the body can follow as a whole. If the body were free to follow both curves from the first dt, then the first thing it would do is warp very badly. Very soon it would be twisted beyond recognition. But real bodies are not free to warp into any shape possible. They already have structure at many levels, and this structure is rigid to one degree or another. So if you try to apply a second circular motion to a real body, you are applying a force that does not just lead to motion—you are applying a force that is trying to break the body itself. It is the molecular bonds themselves that are resisting you. The body does not want to warp. This is why you can apply a second spin to a liquid in circular motion. The liquid does not resist the second orthogonal force. But your second force ends up destroying the "body" of the circular motion, which in a liquid was just a pattern anyway.

That said, it is possible to have simultaneous x and y spins, but you must apply the second spin to a center outside the object. What I mean is that the electron must spin end over end, rather than spin about the axis through its center. To go back to the Earth example, you can see that we can easily imagine the Earth hurtling end over end throughout space, since this end over end motion would not affect its axis spin at all. A gyroscope resists a 90° force, but only because we have fixed the center of the gyroscope relative to the force. A gyroscope will not spin in two ways about its center. But if we put the gyroscope in a spherical container, then we can rotate the gyroscope around a point on the surface of the sphere. We can do this even if the gyroscope is firmly attached to the container. Take a spinning bicycle tire and extend the axle out so that the diameter of the axle is equal to the diameter of the wheel. Attach the ends of this axle firmly to a great sphere with the same diameter, so that the wheel is inside the sphere. You can now rotate that sphere about any point on the surface of the sphere, without the internal motion causing precession. This is because you are no longer attempting to cause two different rotations about the same center. You have created a center just beyond the influence of the first axis.

What is even more interesting is that the circle of this new revolution now has a center that is not stationary—it travels. And it travels in a very interesting way. Let us say you have the

Earth spinning about the x-axis, and you give the center of the Earth a constant velocity in the y-direction. Next, we add an end-over-end spin in this same y-direction. Now, what sort of total curve would this end over end spin create, for the center of the Earth? It would create a wave [you may visit my website, milesmathis.com, to see an animation of this wave motion].

Let that sink in for a few seconds. Albert assumes that both angular momentums are measured about the same center. Beyond that, he assumes that the measured qualities or quantities don't matter. He assumes that angular momentum is conceptually equivalent to velocity or position or any other parameter. He assumes that because that is what all physicists have so far assumed. What matters for QED is how these unanalyzed variables plug into equations. I have just shown that the actual variables matter very much. The whole explanation for QED lies in the real motions of these real bodies, and the explanation is capable of being stated in simple, direct terms, as I did it above. The two angular momentums not only influence each other in specific and distinct ways; the ways they influence each other provide the conceptual and physical groundwork for QED—a groundwork that has so far been ignored.

But let us return to Albert's argument. He gives the electron color and hardness, to simplify the analysis. The electron has four states: black, white, hard, soft. The physicist has equally simple tools. He has a color box and a hardness box. If he feeds in an unknown electron, the color box tells the physicist black or white. The hardness box tells him hard or soft.

Now, if the physicist feeds white or black electrons into a hardness box, half trip the hard detector and half the soft. Likewise for hard or soft electrons fed into a color box. This means, according to Albert, that "the color of an electron apparently entails nothing whatever about its hardness" or the reverse.

The problem encountered by Albert's physicist is that these two simple detectors seem to work in strange ways, if you set them up in combination. If the physicist sets up three boxes like this: color box, hardness box, color box, the percentages at the end are mystifying. The hardness box in the middle is set up so that it captures only one emerging color, which Albert lets be white. The white electrons travel to the middle hardness box, where half of them make it through and go to the last box. The surprise is that of those, only half are white when they come out. Our final color box finds half of them are black. Wow. Albert and QED tell us this is a big problem. It cannot be explained logically. Albert says that his physicist tries everything. He builds his boxes in a variety of ways, to make them more (or even less) precise. It doesn't matter. The same 50/50 split comes out at the end.

This has been one of the central problems of quantum physics from the very beginning. It has been a mystery for at least 80 years. But the outcome is easily explainable once you have my analysis above in hand, regarding the various spins. Let's say you have a sample of electrons and are going to measure angular momentum in both zx and zy planes. If we have four possible outcomes, then we assume that each momentum is either clockwise or counterclockwise, relative to some observer. Now, put yourself in the position of this observer and see what happens. At the first moment, you look and you see that the electron is rotating clockwise about its x-axis, with that axis pointing straight at you. This means that the rotation is in the zy-plane. In other words, you are looking at a little clock, since it is moving relative to you just like the second hand on the face of a clock. That clock face exists in the zy-plane. A moment later the electron has rotated a half-turn, end over end along the x-axis. This rotation is in the zx-plane, about a traveling y-axis. After this half-turn, you look again at the clock face. Its motion is the same, but it now appears counterclockwise to you.

If that was confusing, you can easily perform the above visualization with a desk clock, provided of course that it is not digital. Hold the clock in front of you. Its hands are turning clockwise, and they represent the spin in the x-plane. Now give the entire clock a spin in the y-plane, simply by turning it one half turn end over end. If you do this you will now be looking at the back of the clock. The second hand is now moving counterclockwise, relative to you. It is that simple. That is all I am saying. The second hand of the clock is spinning around an x-axis that is pointed right at you. Then you spun the whole clock around a y-axis. Very elementary, but it shows us that the x-spin of the electron must be variable, if you measure it relative to an observer external to the electron. If the electron has both x-spin and y-spin, then the x-spin will be variable, measured by a stationary device. Only an observer traveling with the electron would measure its spin as consistently CW or CCW. The same thing applies in reverse, of course. If you are measuring the other angular momentum, then you get a periodic variance in the first one.

You could say that the spin changes due to relativity, but that would actually be over-complicating the situation. We don't need any transforms here, and the kind of simple relativity I have just described was known long before Einstein. It is true that my analysis used relativity to find a solution, but it is the simplest, pre-Einstein sort of relativity. It is just to say that an observer must pay attention to how the object he is measuring is changing over time. A measuring device, whether it is an eyeball or an electron detector, is a constant frame of reference, and a spinning electron will show variance with regard to that device at different times, as I have just shown. There is nothing esoteric about it, although I suppose it is a subtle thing to have to notice.

Once we apply this to our measuring devices, whatever they are, we see that this must affect our outcomes quite positively. Let us go inside the first box. It was measuring color, so let us assign color to the clock-face rotation. White is CW, black is CCW. The box finds that some electrons are white and some black. To differentiate, it must apply some field or force to them over some dt. Let us imagine, to simplify, that the box feeds the electrons into a chute, like cattle, and then puts them all through the same door. This door is like the metal detector at the airport, except that it takes a picture of the electron as it rushes through. It has a very fast f-stop, an f-stop of dt. If the electron was CW at that dt, then the box ejects it from the white door. If the electron was CCW at that dt, then the box ejects it from the black door.

This is, in fact, very much like the way detectors work. They don't take pictures, of course, but some sort of force or field separates the white and black electrons. The field may not be limited to a dt, but the first impression of the field is crucial. The electrons are moving quite fast, and the time periods are therefore quite small. The field doesn't have time to snap a bunch of pictures and start changing its mind.

What this all means is that whiteness and blackness and softness and hardness are not constants. Every electron is both black and white and hard and soft, at different times. But it is all those things only if you sum over some extended period of time. At each dt, it is either hard or soft, black or white. It is not both at the same time. At one measurement, it will be one or the other. Over a series of measurements, it will be both.

This is the subtlety that QED has never penetrated. It explains the above problem like this: If you put electrons like those I have described through a color box, the color box sees some of them as black and some as white over the dt measured. But they are actually not white or black as they come out—they remain potentially both, depending on the point in the wave you measure. If you measured the white ones coming out at a different point in the wave motion, you would find them black, and vice versa. Now, the color determination is repeatable, since a similar box will catch the electrons in similar ways. All color boxes tend to shute and channel electrons in the same way, so that the exiting group is made coherent. A second color box must then read them the same way as the first.

What happens in the second box (the hardness box) solves the mystery. The second box creates coherence in the second angular momentum. This assures that other hardness boxes will find the same hardness. But in creating this coherence, the second box re-randomizes the first variable. Why does it do this? It does this because the wavelength of the two angular momentums is different. If the first wavelength was taken as R, for the radius of the electron, then we have to take the second wavelength as 2R, for the diameter. This is simply because the second wavelength is caused by end over end rotation. If we cohere the end over end rotation, this must split the measurement of the axial rotation. If we cohere the axial rotation, this must split the measurement of the end over end rotation. One is half the other, so you cannot create coherence in both at the same time.

I can show this with simple waves in two dimensions. Study the diagram below. We have two opposite combinations of ½ and 1 waves. If you synchronize the ½ waves, the 1 waves are off. If you synchronize the 1 waves, then the ½ waves are off. You cannot synchronize both. This, in essence, is what is happening in box two. The hardness waves are being made coherent, so that the color waves are being thrown out of synch. The third box then reads them as ½ one and ½ the other.

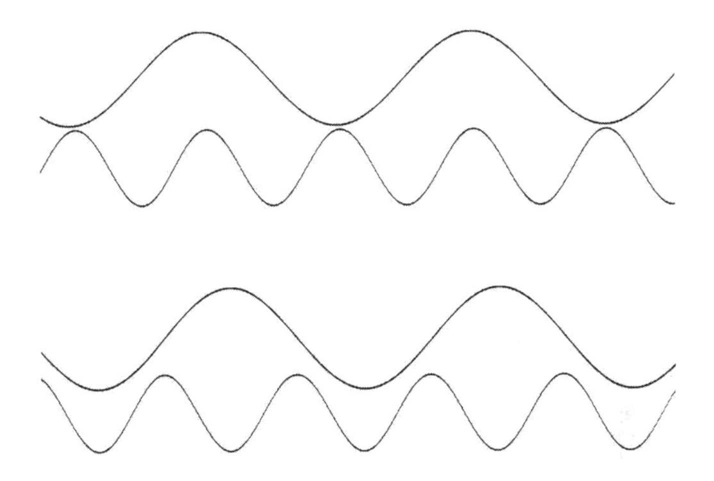

You can see that I have simultaneously solved the problem of superposition and the problem of the wave motion of quantum particles. I did this simply by noticing that the second angular momentum must be about a center that is just external to the object. That is to say, the y-spin is end over end.

With the hindsight this gives me, it seems shocking that this was not seen earlier. The reason it was not seen is that Heisenberg and Bohr convinced everyone early on that Quantum Mechanics could not be explained with straightforward logic and simple visualizations. No one has ever bothered to apply a little commonsense to the physical situation. They were so sure that it couldn't be done, that they didn't even try to tackle the problem on a visual or mechanical basis. This predicament soon snowballed, since as more and more great physicists looked at the problem and failed to explain it, later physicists became more and more sure that it couldn't be solved. They did not want to waste their time combing something that every genius from

Bohr to Feynman had already combed. That seemed not just foolish, but sacrilegious. But the fact is that there has probably been no one since Bohr that tried very hard to make classical sense of the problem. Physicists who came right after Bohr took his word for it, and contemporary physicists have reached the point where most don't even want a mechanical explanation of QED. The spooky paradoxes are more fun. They make better copy.

You may now go to my second paper on superposition, to see a similar experiment solved even more quickly and transparently. That experiment is the famous one of two beam splitters and two mirrors. In that paper I also offer three more diagrams, which may be helpful to many.

I think it is obvious that the end over end spin in the y-direction can be applied to other problems, including the propagation of photons, the two-slit experiment, and so on. In subsequent papers I will apply my finding to the electron and proton and to a large list of mesons, to show that the same four stacked spins can explain all quantum make-up and motion. I will also have a lot more to say about other specific problems within QED and QCD, and their solution with straightforward logical analysis.

SUPERPOSITION AGAIN

I recently received a question from a reader, concerning my superposition paper:

I've studied your paper and it made sense to me. However, there are other cases of superposition that are available that do not require measurement of x,y spins of electrons. For example this video at youtube* shows a well-known experiment of superposition using beam-splitters and mirrors. Although I'm personally skeptical of "same particle in multiple places at the same time" argument, I'm struggling to come up with a better explanation. So here is my question: how would you explain this particular superposition experiment? Is it really the same photon in multiple places?

What this reader is talking about is an experiment where a light beam is split, mirrored symmetrically, then split again (see diagram below). Detectors are set up at the second split, and we have another big and spooky "mystery". The magicians at youtube claim this is because the same photon goes both paths and interferes with itself.

This is a similar question to the one I solved in the first paper, but here the magicians vary the setup to fool the audience. When I say "magicians" I am not just being sardonic. This really is a case of prestidigitation, like a shell game. The fake physicists misdirect your eye, and by the end of it you can't say where the photon is or why it is there. Just as almost no one, no matter how smart they are, can tell what the trick is in a good magic trick, almost no one can sort through all the misdirection and fast talk of the current superposition patter.

First of all, to solve this, we don't have to "measure" spins of electrons or photons. We just have to give the particles wavelengths, and the standard model already does this. The magicians at youtube do this, and they even try to prove double existence by manipulating wavelengths (in a very sloppy manner). But they don't define the wavelengths carefully enough. This forces them to solve the problem by proposing the impossible. It does not bother them to propose the impossible: in fact, they enjoy being magicians. They enjoy performing miracles and dumbfounding the audience.

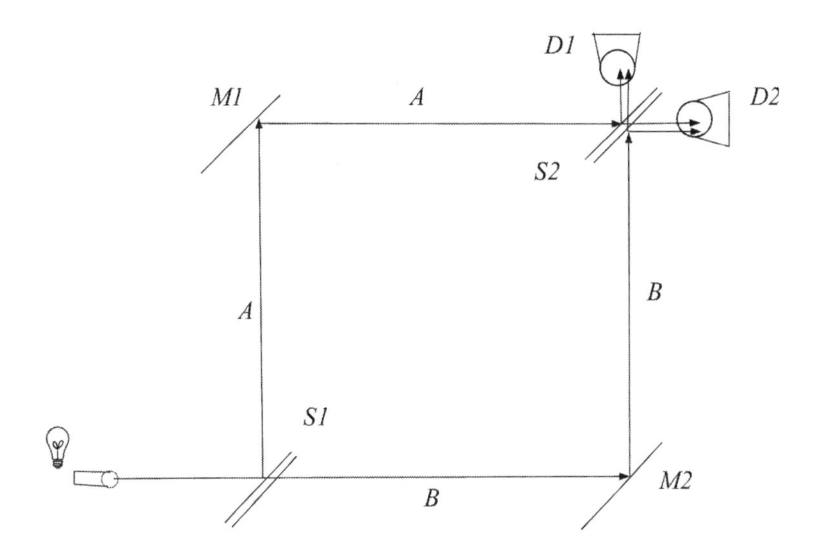

But let's solve it with mechanics instead. We know that the first beam splitter S1 splits 50/50, since if we move the detectors up to S1, the detectors tell us this directly. The second splitter S2 is exactly like the first, so we should expect the detectors at the end to give us the same 50/50 split (we are told). Instead we find all the photons at D2. Big mystery.

At youtube they explain it in this way. If we fire the photons one at a time, the photon takes both paths and interferes with itself, keeping it from reaching D1. The problem with this answer is not just that the same photon travels both paths, although you would think that would be enough to disqualify the answer. The other problem is that if the single photon has interfered with itself, how does it reach D2? We have a detection at D2, remember? The standard answer is that the interference only happens with the half of the photons that pass straight through the splitter on path B. The half that are split are turned directly into D2.

So, if a photon on path B passes straight through S2, it interferes with itself, and doesn't go to D1. If a photon on path B is turned, it doesn't need to interfere with itself, and it goes to D2. If a photon on path A is going to be turned at S2, it interferes with itself and does not go to D1. If it goes straight through S2, it does not interfere with itself, and goes into D2. That is the magic answer.

Not only is that answer much more complex than it needs to be, it is contradictory. Along path A, the interference takes place on the near side of the splitter. The photon on the A path does not go straight through the splitter: it waits for its twin to go through the splitter on path B, and only then is the interference completed. But if the photon is moving on path B, it goes through the splitter and then interferes with itself. The interference takes place on the A side of the splitter both times. Not only are the paths not symmetrical, there is no way to explain how the photons know whether they are the primary photons or the twins. In other words, the youtube magicians haven't explained why the interference always takes place on the A side of S2. Why doesn't the interference ever take place on the B side of S2, after the photon on path A has passed straight through S2?

Also, you can see that they need the single photon to take both paths every time, just in case it is needed. This is what the sum-over proposition of Feynman means. Every photon takes every possible path, then we do the math at the end, to cancel wavelengths and decide where particles will be detected. But if that is the case, why aren't the twin particles detected when the detectors are at S1? In other words, once they explain the action of the splitter and photons at S2, they have to go back and see if it works at S1. We have the proposal that all photons take both paths. If they are on both paths, why did the detectors at S1 find a 50/50 split? Why do detectors detect primary particles but not twins?

This explanation wants the photon to take both paths in the second case, where the detectors are at S2, but it doesn't want the photon to take both paths when the detectors are at S1. If the photon is on both paths, then both detectors at S1 should detect all the photons. Yes, logically, we should detect 100% more photons than are emitted, since we would be detecting both the particles and their twins.

So the current magical explanation not only wants us to believe that the photon takes both paths, it wants us to believe it is on the path and not on the path. It is on the path when we want it there to interfere with itself, but it is not on the path when we don't need it to interfere. The current explanation is not one miracle, it is two miracles stacked.

The funny part is that the youtube magicians tell you the right answer, but then deflect you from noticing it is sufficient, without interference. They admit that each turning will shift the wave ¼ wavelength. If the wave passes straight through a splitter, it is not shifted. So, in order to reach D1, the wave is either shifted three times on path A, or one time on path B. To reach D2, the wave is shifted 2 times on either path. This tells us immediately that the experiment prefers even shifts. We should then seek to explain this without interference or doubled particles.

The splitter, that we expected to work the same way in both positions, is not working the same way in both positions. At S1, it is letting half the particles pass straight through. At S2, it is letting all the particles on path A pass and none of the particles on path B. Why?

The answer is even simpler than my answer to the detectors-in-sequence problem of my first paper. As in that paper, the first splitter is acting as a polarizer. It is sorting the photons coming from the emitter. All the photons going on path A have the same orientation, and the same for B. They are on the path they are on because they reacted the same to the material in S1. The photons on path A are all equivalent in orientation to each other, but they are opposite in orientation to the photons on path B.

This means the splitter at S1 is dealing with a different incoming group than the splitter at S2, and we should not expect the splitter to act the same in the two places. The first problem, therefore, is our expectation that they should act the same. The magicians tell us that the logical expectation is that the splitter should act the same in both places, but that is false. It is either a lie or a very big and obvious mistake.

The splitter at S1 is receiving one group of mixed photons, from one direction. The splitter at S2 is receiving two groups of polarized photons, from two directions, and each group is opposite the other group.

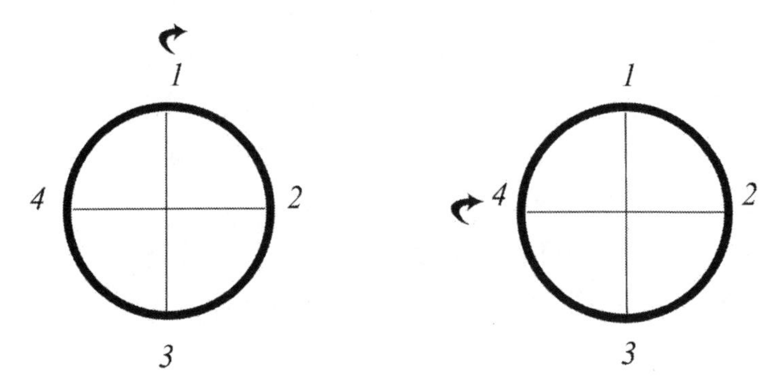

Let us show this in more detail, but still very simply. Let us say photons can either be spinning around a vertical axis or a horizontal axis, relative to the first splitter. In other words, if we simplify the photon into a circle, it is either spinning along a 1-3 axis or a 2-4 axis. All our emitted photons are either 1-3 or

2-4. If they are 2-4, the splitter lets them pass straight through along path B, without deflection. If they are 1-3, the splitter deflects them along path A. But in deflecting them, the splitter turns them ¼ turn, as the magicians on youtube tell us as they read from the internet. This means that the number 2 is leading on both paths. When the particles are turned by the mirrors, they each shift ¼ turn, so that the number 4 is then leading on both paths. The mirrors are opposite in orientation themselves, so we turn the B particle clockwise but the A particle counter-clockwise. But on path A, the particle is still spinning on the 1 axis, and on path B, the particle is still spinning on the 2 axis. So the particles approach the splitter at S2 as shown in the diagram

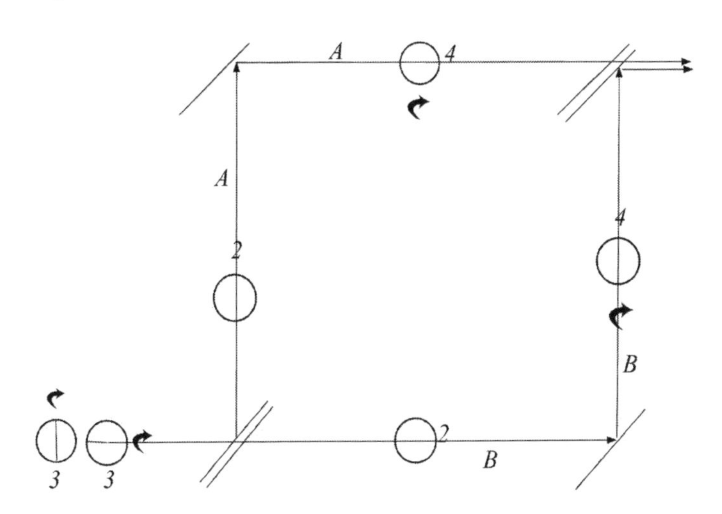

The particles on both paths are now reversed from their original orientations, as you can see. So the splitter reacts to them in the opposite way, turning the B particle and letting the A particle pass.

All the particles on A are the same, so the splitter reads them the same way, letting all of them pass. All the particles on B are the same, so the splitter reads them the same way, turning all of them. Very simple. Not mysterious at all.

Not only is there a mechanical explanation, the explanation is quite quick and transparent, yielding to very simple diagrams. We don't need any interference or doubled particles or multiple paths. The youtube video tells us that the only way to explain loss of detection at D1 is by interference, but I have just shown that is false. The expert on the video also tells you to trust him, but that is very bad advice. Never trust anyone, least of all a scientist. Science is not about trust, it is about a logical and physical explanation.

Of course, this once again destroys the Copenhagen Interpretation and 90 years of physics. Quantum physicists have been assuring us that this couldn't be done. They have assured us that no logical answer could be given, and that no diagrams could be drawn. I have given them and drawn them, as you can see with your own eyes.

Some will complain that my explanation requires spin, whereas the current theory gives the photon a wave, not a spin. My answer is that it doesn't matter one way or the other. I believe the photon is spinning, and have shown theoretical and physical proof of it elsewhere, but my explanation here doesn't require you to believe it. The spin in this explanation simply allows me to show the wave more easily. The standard explanation of superposition comes from Feynman, and it is likely these youtube people are reading something by Feynman off the internet as they make their film. Well, Feynman also invented a thing called the shrink-and-turn method, which I pull apart in another paper. To illustrate the wave, Feynman uses little clocks, much like I have here. That is, he draws a circle with numbers on it and lets that stand for the wave as the photon travels. He doesn't call it a wave, true, but it works just like my wave here. His method works precisely because it mirrors my mechanics here. Well, take the little circles above as waves if you like, rather than spins. Spins create waves in a direct manner, so they are great for illutrating waves even if you don't like spins. If you don't want to assign the waves to spins, fine with me. Assign them to wobbles or leaps or hiccups or to nothing. I don't care. The point is, I solved the problem with diagrams, mechanically, without interference, without ghost particles, without multiple paths, without spooky forces, and without mystification or magic.

And finally, as a bonus, I give you the fact that the current explanation of superposition, using light interfering with itself, contradicts the current explanation of the Sagnac Effect. Wikipedia admits that the Sagnac interference math is the same both before and after Relativity. Classical physics made the same predictions as post-classical physics, regarding this effect. And, since the Sagnac Effect already had a satisfactory explanation and math before quantum physics, it didn't require the sort of explanations that have been devised for superposition. This despite the fact that the two experiments have much in common, as you see, using mirrors and beam splitters (a half-silvered mirror is a sort of splitter) and square circuits. The reason this contradicts the Sagnac Effect is that, to be consistent, we have to take the quantum explanation into that experiment as well. We can't have light interfering with itself in some cases and not interfering with itself in other similar cases, just to suit sloppy theorists. If light takes all possible paths, why doesn't it do so in the Sagnac experiment? If we let light take both paths in the Sagnac experiment, we immediately ruin our math and our explanation. Instead of getting light where we need it, we get light where we don't need it. We have too much light on both paths, and the result is either a total cancellation or a big mess. This is the problem with so many of the current jerry-rigged theories: they are very problem specific, and the magicians just hope you don't try to universalize them, and apply them to similar problems. Because if you do, you find out that they are completely ad hoc, and therefore physically false.

*http://www.youtube.com/watch?v=qpQABLRCU_0

A MECHANICAL EXPLANATION OF ENTANGLEMENT

In March 2009 *Scientific American* ran an article called "Was Einstein Wrong?" by David Z. Albert and Rivka Galchen. Albert is in the philosophy department and Galchen is in the English department, both at Columbia. Not too long ago, Feynman was ridiculing "physics by philosophers," and being cheered for it across the country; but now, less than two decades later, we have not only philosophers being published as physicists, we have teachers of creative writing being published as physicists. Now, I have nothing against philosophers or English teachers: I am simply pointing out the hypocrisy of it. Philosophers are not allowed to have opinions about physics unless they teach at some place like Columbia, have been fully indoctrinated (at Rockefeller University, no less), and sign on completely to the current disinformation campaign. In other words, *your* philosophers cannot have opinions on science; but *ours* can. We will not publish philosophers who disagree with us, but if a philosopher is willing to parrot our views entirely, we will.

I have run across Albert before. You will remember that I critiqued his book *Quantum Mechanics and Experience* two chapters ago, in my analysis of superposition. I solved that problem for him in a simple mechanical way, with diagrams, and sent the solution to him directly. But he prefers to stay with the "in" party rather than be correct. He did not reply to my paper, and has now switched mysteries. He is now concentrating on entanglement, the subject of this paper at *SA*. In a nutshell, Albert has accepted the interpretation of John Bell that entanglement must imply non-locality. In one short paragraph, Albert poorly glosses Bell's argument, and then states, "And so the actual physical world is non-local. Period."

Here is Albert's entire argument for that final decision: "if no algorithm could avoid non-localities, then they must be genuine physical phenomena. Bell then analyzed a specific entanglement scenario and concluded that no such local algorithm was mathematically possible."

Yes, well. Albert claims that Bell said that no algorithm could avoid non-localities. In direct language, that means that Albert thinks that Bell said no mechanical explanation of entanglement was possible, or, conversely, that a mechanical explanation was *impossible*. I would say that interpretation of Bell is pretty strong, but let us say it is true that Bell meant that. Can Bell be correct? No. Albert must be a very poorly trained philosopher if he believes you can prove a negative in this form. Every good philosopher since Thales has known that you can't prove a general existential negative. Yes, you can prove that a theorem is false, but you can't mathematically or logically prove that something cannot be done, universally.

This should be doubly obvious in Quantum Mechanics, the realm of probabilities. Any first-year statistician knows you cannot prove anything with probability math. But Albert expects us to believe you can prove a negative with probability math and probability assumptions.

The reason you cannot prove a universal negative is that it requires total knowledge of the field. To make the claim Bell and Albert are making would require them to know, for a certainty, that they knew everything about the mechanics, operations, and interactions of the quantum field. Norman Finkelstein has entitled one of his books *Beyond Chutzpah*, and this argument of Albert's is beyond chutzpah. It is hubris, period.

To give you an example, 180 years after Faraday and 230 years after Franklin, we still don't know how the charge field works mechanically. Today, top physicists will tell you the charge field is mediated by messenger or virtual photons, photons that are able to "tell" the quanta "move closer" or "move away." Yes, the same photon can cause negative or positive charge. As a matter of science, that is clearly infantile. I have proposed a mechanical charge field that is always repulsive, and this begins to answer some of the mysteries of QED. But, regardless of any of my theories, which are admittedly in the first stages, the standard model is nowhere near omniscience about the charge field. And **it is this charge field that must mediate entanglement**.

My assumption is that entanglement will be explained by some simple mechanics. In fact, I am quite near to that solution myself (very close: only a few minutes away: see below). This was also the assumption of Einstein. In the famous EPR paper, Einstein argued that entanglement was proof that quantum mechanics was incomplete. Bohr disagreed, but he was simply saving face. He never argued for actual non-locality, as Albert admits. He only argued that quantum mechanics was as good as it was going to get.

Amazingly, contemporary physics is finally moving past the Copenhagen interpretation. Albert says in his paper that this is happening and that it is of historical importance, and I agree with him. The publication of his paper is proof of that by itself. Albert is dismissing both Einstein and Bohr here, which would have been *verboten* in the mainstream until very recently. But the standard model is moving past the Copenhagen interpretation to become less mechanical, not more. The Copenhagen interpretation, which was already a flight into the non-physical, was still too restrictive for contemporary physics. Bohr's philosophy was already pseudo-philosophy, an enshrinement of irrationality, but it was not irrational enough to suit the contemporary physicists. The contemporary physicist wants no limits on his ability to perform mathematical magic, so the idea of locality must be given up.

This is revolutionary, in the worst way, in that it removes the last rule of logic and theorizing. Bohr had already destroyed mechanics, which was bad enough. Planck, Schrodinger, and Einstein never forgave him for it. But even Bohr had been forced to admit that there was probably something physical going on beneath the wave function. Bohr believed it was forever hidden from us, but it was there. It had to be.

But with Albert's interpretation of Bell, all that is gone. The dam has broken completely free, the last rock is dislodged, and theorists in the future will not have to obey any rules except the ones they make up freely.

In fact, physics will no longer be physics, since it will no longer be physical. It will be entirely mathematical. *Scientific American* will have to change its name to *Unscientific American*, or *Heuristic American*, or *Magician's Monthly*.

I have called this argument of Albert's disinformation, and some might say that is unnecessarily incendiary. But if you read this paper at *SA*, you will see that it reads like propaganda. Albert is not only selling a viewpoint, he is trying to sell that viewpoint by dressing it up as its opposite. That is disinformation and agitprop, by definition. He says that after the Copenhagen interpretation,

> To spend any more time on these matters became, thereafter, apostasy. The physics community thus turned away from its old aspirations to uncover what the world is really like and for a long time thereafter it relegated metaphysical questions to the literature of fantasy.... From the early 1980s onward, the grip of Bohr's conviction—that there could be no old-fashioned, philosophically realistic account of the subatomic world—was everywhere palpably beginning to weaken.... The old aspirations of physics to be a guide to metaphysics, to tell us literally and straightforwardly how the world actually is—aspirations that had lain dormant and neglected for more than 50 years—began, slowly, to reawaken.

That, my friends, is purposeful misdirection. Albert tells us outright that the Copenhagen interpretation is being bypassed in order to "tell us literally and straightforwardly how the world actually is," to give us an "old-fashioned, philosophically realistic account of the subatomic world." When the exact opposite is true: the Copenhagen interpretation is being bypassed to give physicists room to propose even more unrealistic, non-literal, non-straightforward, and non-old-fashioned theories. Non-locality is not realistic, is not literal, is not straightforward, and is not old-fashioned. Albert is only using those words to fool you into accepting something you would not think of accepting dressed in its own garb.

Now let me show you some of the ways that entanglement is misinterpreted. If we go to Wikipedia for the modern gloss, we find this:

Quantum mechanics holds that states such as spin are indeterminate until such time as some physical intervention is made to measure the spin of the object in question. It is equally likely that any given particle will be observed to be spin-up as that it will be spin-down. Measuring any number of particles will result in an unpredictable series of measures that will tend more and more closely to half up and half down. However, if this experiment is done with entangled particles the results are quite different. When two members of an entangled pair are measured, one will always be spin-up and the other will be spin-down. The distance between the two particles is irrelevant.

A close reading of those few sentences already shows how the mystery of entanglement is a manufactured mystery, created by false probability assumptions. The problem in this Wiki quote is closely related to Schrödinger's cat mystery. In a thought problem, Schrödinger put a cat in a box and then assigned a probability number to the cat: say, .5 the cat was alive, .5 the cat was dead. We can't see the cat, so we don't know. Quantum mechanics says the numbers are all we know. Schrödinger says no, there is some fact underneath the numbers: either the cat is dead or it is alive. When we open the box, it must be one or the other, not both. Amazingly, the princes of QM did not say, "Yes, well of course. But we don't know until we open the box." That would have been sensible. No, QM said to Schrödinger that the cat was NOT really alive or dead. It was neither alive nor dead until we opened the box and saw it!

Yes, that is the level of philosophical understanding of modern physicists. Schrödinger *lost* that argument, which is why I still have silly things like that that I can quote from Wiki.

Contemporary physicists actually believe that the "physical intervention" of measurement determines part of the math. It does this via the HUP, the Heisenberg Uncertainty Principle. A certain interpretation of the HUP makes the physicist an actual part of every equation, and this interpretation is now the accepted one. That is spooky enough, in itself, but entanglement is even spookier. Using the anti-Schrödinger interpretation of cats, QM had decided that nothing could be known about particles except their probabilities. In other words, there was no certain knowledge beneath the numbers. But with entanglement, we get certain knowledge from probabilistic situations. With entangled particles, "one will always be spin up and the other will be spin down." Note the word *always*. That is certain knowledge.

To explain this, quantum physicists have come up with the idea that the particles are in contact with each other over huge distances, without any mediating field or particle. Yes, they can talk to each other instantly, so that when the physicist measures one as spin up, the other can flip immediately to spin down to conserve parity.

All this is patently absurd, but neither the physicists nor the philosophers can seem to cut through to the fairly obvious answer. They can't do that because they have made the question much more complex than it is. First of all, the physicists have buried the problem under decades of math and terms. Then the philosophers have followed, adding their mountain of semantics and lingo and sloppy thinking. One must come to the conclusion that neither the physicists nor the philosophers *want* a simple answer. They only want to look smart, bandying a vast vocabulary and an infinite disrespect for their readers.

The fairly obvious answer is that their first postulate was wrong. They assumed that there was no reality under the probability numbers, but entanglement showed that there was. Just look at the Wiki quote again: the whole problem is between their postulate and the outcome of the experiment. Faced with a contrary experimental outcome, a sensible person would admit his postulate was wrong, but that is not the way of modern physics. Physicists cannot admit they were wrong. So, in order to keep their postulate, they stoop to this force-at-a-distance magic.

Albert even admits that non-locality is force at a distance, which puts physicists right back with Newton. In order to keep their "modern" edgy pseudo-philosophical postulate, they are willing to turn the clock back on the entire field 300 years. This is the only way that superseding the Copenhagen interpretation can be seen as "old-fashioned": it takes us back to the time of the Inquisition, before anyone was capable of doing mechanics. If we accept non-locality, we can wipe out all of physics since Galileo, and we can wipe it out in the name of "progress," as Albert does in this paper. As art is now post-modern, physics is now post-mechanical. Physics is post-physical.

Ironically, Feynman did not believe in this force-at-a-distance dodge. I have been hard on Feynman in other papers, and he continued a lot of the propaganda of Bohr and Pauli; but not all of it. If you study my explanation of his shrink-and-turn method in a recent paper, you will see why. I will gloss it here to explain entanglement mechanically. In his book *QED*, Feynman explains partial reflection by glass by assigning an arrow and a clock to each individual photon. I have shown that the arrow is a vector and the clock is a spin. The clock and arrow, taken together, are able to tell us where in its spin cycle each photon is. In other words, we can calculate where the particle is in *its own* wave.

Now, Feynman was never able to give this mechanical interpretation; or if he was, he never admitted it. But his method would have allowed him to explain entanglement without force at a distance. In this way: since each photon has both a turning clock and a vector, each photon has both a wave motion and a linear motion. This means that the wave belongs to each photon, not to the set of photons. This is revolutionary because, in this way, light is no longer analogous to sound: it is not a field wave, but a particle with spin. The wave belongs to *each* particle, and may be assigned to a mechanical motion: spin. If each photon has a real spin with a real wavelength and a real period of rotation, then we can use that period of rotation to track it. Using Feynman's little turning clock, we can follow the photon, no matter how far it travels, and predict with some certainty what state it will be in. We cannot say that the clock will be at 6 or 12, but, given an initial state, we can predict a final state. *If* the clock was initially at 12, after some time we can predict that the clock will be at 12 again. To do this, we only need to know the period of rotation and the time of travel. If we know the wavelength, we can calculate the period of rotation quite easily, so this is not a difficult problem mathematically. Once we sort through the mechanics, the math becomes simple.

This explains entanglement because we do know an initial state. We don't know if the quanta are at 12 or 6 on the clock face, but we might know one is opposite the other, for example. If one is at 6, the other is at 12. If they have the same periods of rotation, then after any time, they will still be opposite, without any communication between them. Other relationships will also be trackable and stable, as long as the periods of rotation are known relative to each other. In other words, as long as we know sizes and wavelengths, we can predict comparative wave positions at any distance or time away from collision.

This is the mechanical explanation of entanglement, without spooky forces. Albert and Bell have both been proven wrong, by direct demonstration.

THE DOUBLE-SLIT EXPERIMENT

In this paper I will show the simple mechanical solution to the famous double-slit (or two-slit) experiment. Feynman called it the most important experiment for understanding quantum motion, and he may have been right. The most difficult problems are always the most important, and this one has remained unsolved up to this minute. Thomas Young first performed it in 1801, which means it has been a mystery for over 200 years. Even Feynman failed to solve it. He offered a mathematical solution only, but was not able to provide a physical solution.

I will solve the two biggest problems here: the problem of the single photon and the problem of the detector. In the first problem, we let photons go through the experiment one at a time. Using the photon-as-particle theory that Einstein proved and Feynman confirmed, we expect no wave interference, since the photon must go through one slit or the other. But we see interference. The single photon seems to be interfering with itself in some strange way. Up until now, there have been several proposed solutions. The first is that the "wave front" of the photon goes through both slits and interferes with itself. Since the wave front is still not defined mechanically, this solution is not very compelling. The wave front is and always has been defined using Huygen's visualization. The wave is seen as a semi-circular forward transmission from every point on a line of moving photons. With a single photon, this would be a semi-circle in front of the photon. But we are never told how far this semi-circle extends, what it is composed of, or how it acts upon the field. So we will let this explanation pass as wholly unsatisfactory.

Another solution is to define the photon as a probability. A discrete particle cannot go through both slits at once, but a particle as probability can (as long as we define probabilities in certain ways).

The third and currently accepted explanation is an extension of this second one. Feynman proposed that the photon-as-probability traveled *every possible* path, and therefore through both slits. Each path is given an equation, and we "sum over" all these equations. If we sum over in the correct way, we achieve interference.

Now, admittedly this is a clever mathematical solution. Feynman was a master of clever mathematical solutions, and this is one of his best. Mathematically it works. But it is not a physical or mechanical solution. It is a mathematical solution. Feynman was not so much a physicist as he was a mathematician that had invaded the physics department (the same could be said of most modern physicists). In his own way, Feynman admitted this. He did not admit to being an invader, but he admitted that his solution was only mathematical. He knew as well as anyone that it wasn't physical, by the old definition of physics. He got around this by claiming that new physics was and must be mathematical only, since there was no possible mechanical solution.

He was wrong, as I will prove very quickly. His math works precisely because there is a physical reality underlying his probabilities. Probabilities are not the *causa sui*, the cause of themselves. It is illogical—even as a piece of mathematics—to propose that probabilities are spontaneously generated, or that they are primary generators. No, they must be generated by a real field. Even in pure mathematics, probabilities are always secondary numbers, produced by an underlying field of numbers. A field of probabilities cannot be a foundational field. They therefore cannot take the place of a physical field.

What foundational field of numbers is creating the probabilities in this experiment? The answer is: the foundational E/M field. In every analysis of this problem and this experiment to date, the analysts have over-simplified the problem. They have assumed, without even putting the assumption into words, that the experiment is taking place in a sort of void or vacuum. The only things they look at are the slits and the photons. But the slits and the photons are not the only important players in this field. Even if you ran this experiment in a vacuum, with the walls and the photons as the only objects in the vacuum chamber, you would still not have a void, since the walls are still material objects. As such, they must be emitting an E/M field. The wall, even in a vacuum, is radiating a field all the time. It is this field that the photon must move through.

I have shown in previous chapters that the charge field, if defined mechanically, must have mass equivalence. If it has mass equivalence, it must have materiality. In other words, the field that mediates the charge between proton and electron must be made up of discrete particles itself. What is now called the messenger photon cannot be a virtual particle with no mass or energy. It must be a real particle and create a real field. I have already given this charge photon a new name (the *B*-photon) and a definite radius (G times the hydrogen radius), so I feel very qualified to use this particle to explain the two-slit experiment. I have not dreamed up this field as a virtual field, a summed-over field, or an *ad hoc* field; I have shown the physical and mathematical necessity of it, and its place in Newton's gravitational equation.

This being so, we must now recognize that our central wall in the two-slit experiment must be radiating this field (I am talking about the wall which contains the two slits). Our single photon must be moving through this field emitted by the central wall. This changes everything in regard to the experiment. The first thing to notice is that we have interference patterns set up by the slits even before the single photon is emitted. If we know that every atom in the wall is emitting this field, as a simple bombarding field, the two slits will create an interference pattern in the field without a single particle moving through the field. *The interference patterns are already there!* The single photon does not create them. The probabilities do not create them. Karl Popper's "propensities" do not create them. **The real atoms in the wall create the interference pattern, with simple spherical emission.**

The only problem is that we cannot "see" this field. It does not create any lines in the far wall, since *B*-photons are not the same size or energy as the single photon we put through the device. Our far wall is chosen because it is made of a material that reacts when the single photons (or electrons, or whatever particles we are using) hit it. But it does not react to the foundational E/M field. It does not react to *B*-photon radiation. This field therefore remains invisible to us. We don't "see" the interference patterns until a large enough particle moves through the field.

The motion of this particle through the field and its reaction to the far wall give us our only data. The experiment is not set up to give us any data about the *B*-photon field, except indirectly.

This simple mechanical explanation not only solves the single photon problem, it also shows why different particles are affected in different ways by the same field. It is quite easy to see that an electron will be funneled by this *B*-photon field in a different manner than a photon, due only to the size difference. If the photon is like a baseball moving through a field of golfballs, the electron will be like a bowling ball moving through a field of golfballs. Put simply, the electron will be funneled much less efficiently. It will resist the field more successfully, and the field will be upset by its presence to a greater degree. All this will now be visualizable, predictable, and mechanically transparent, due only to the discovery of the pre-existing interference pattern and the real field that creates it.

Now let us look at the mystery of the detector. It has been found that putting a counter or detector in either slit changes the entire data in ways that are not predictable with current mechanics. Specifically, a detector in one slit will destroy the entire interference pattern, returning us to a single pattern on the far wall. The current explanations for this are even more tenuous than explanations for the single photon, since Feynman's sum-over trick does not explain it. Attempts to fudge an answer by claiming that we must now sum-over both before and after the detector don't answer the problem unless it is shown how the detector changes the total path. No one has yet done this. Unless Feynman can show *why* AB + BC is not equal to AC (with the detector at point B), he cannot develop an equation for AC that is different from the case without the detector, and therefore cannot show a sum-over variance. Feynman never claimed to any *physical* knowledge of the two slit experiment, therefore his math does nothing to solve the detector mystery.

The detector mystery has led to even more absurd solutions than the single photon mystery. Along with entanglement, the detector mystery has been one of the primary causes of neo-idealism in physics. Many physicists now believe that simply wanting to know something changes the entire experimental set-up, as if asking a question can physically interrupt a field. In this way, physics has crossed over into mysticism. For what I will show are the flimsiest of reasons, physics has chosen to accept spooky forces and psychical interference in their experiments. Rather than continue to look for mechanical explanations, they have preferred to be satisfied with magic.

But again, the real solution is simple and logical and mechanical. The detector is a device with real size and materiality. It inhabits space in or near the slit. This detector creates a real field of its own. If it didn't, it couldn't detect anything. Wikipedia says,

> The detection of a photon involves a physical interaction between the photon and the detector of the sort that physically changes the detector. (If nothing changed in the detector, it would not detect anything.)

Logically this is true, but it fails to describe the correct interaction. It is not the interaction between the photon and the detector that deletes the interference pattern, it is the interaction between the *field* created by the wall and the *field* created by the detector. The field of the wall is a *B*-photon field, which is made up of exceedingly tiny particles. The field of the detector, whatever it is, must trump this field. We do not have detectors that make use of the *B*-photon field, since, up to now, we haven't been aware of it. All our detectors use "larger" fields, since these are the only fields we can create and use. To use the ball analogy again: if the *B*-photon field is a field mediated by golfballs, the field used by our detector to detect the particle passing through the slit must be made up of baseballs or bowling balls. This detector field is obviously going to destroy the golfball field and all patterns in it. Our particle passing through the slit is then going to be funneled by the detector field only. This detector field doesn't create an interference field, so we do not see one.

I will now make a prediction that will prove that my solution is correct. If we house our particle projector in a wall like the far wall, one that allows us to mark a hit by a photon or other particle, we can monitor particles that are reflected by the central wall as well as particles that pass. As for the central wall, we will make it reflective on the side near the projector (without changing its makeup in any other way). We simply want to be sure that, in the case our particle does not go through one of the slits, it bounces back and returns to the first wall. We don't want the central wall to absorb our particle. OK, now for the prediction. If we purposely fire our particle so that it fails to go through either slit, and it hits the central wall and reflects, I predict that we will find the same interference pattern on the near wall that we found on the far wall. Given the current explanation of Feynman and others, there is no way this could be true. We have no particles going through slits, therefore no sum-over solution will explain the interference. But my solution accounts for it in a very straightforward way. In my solution, it is the central wall that is creating the initial interference pattern, and due to the position of the central wall, it must be creating the same pattern both backwards and forwards. This being true, we must find a very similar pattern created on the near wall to what we found on the far wall. In this case, we will have the appearance of interference with NO particle going through a slit. Current theory thinks it is mysterious that one particle going through a slit creates interference, but this experiment will show interference with no particle going through a slit. This is because the slits are creating the interference on the near wall as well as the far wall.

You will say that the first and third walls must also be emitting my B-photon field, supposing it exists, and that this must skew my solution. But this is not true. The near and far walls are emitting that field in a linear way, since they do not contain any gaps. Only the central wall is emitting an interfering field, due to its shape. Since all the emitted fields are real, they must interfere with each other in some way, and to some extent. But since the far wall is emitting a rectilinear field, or its equivalent, it will not change the shape of the fields from the central wall. It may tamp them down a bit, as a matter of total energy, but it would not be expected to destroy the curves.

The only real problem in the set-up I have described is the hole in the first wall created by the projector itself. This hole will create a ripple in the B-photon field emitted by that wall. I think this factor could be removed from the experiment by setting our projector above the first wall, instead of in it. Then all we have to do is calculate the proper reflection angle, so that particles reflected from the second wall are sure to hit the first wall. B-photon emission from the body of the projector itself will still disturb the fields set up by the walls, but perhaps this can be minimized in other ways. Even if the effect of the projector cannot be removed from the experiment, we will still find patterns created on the near side that cannot be explained by Feynman or the current model. Even if the first wall continues to emit like a single-slit wall, no matter what we do, we will still find the near field acting like the sum of a single wave meeting a double wave. My method allows us to calculate and predict this field. The current solution cannot predict or explain these patterns at all, since it doesn't recognize that the patterns are already there before any particle is fired.

You can see that once again a simple mechanical explanation has utterly destroyed decades of murky and muddy hypotheses. An entire sub-field of physics has been destroyed with a few pages of elementary logic. And the entire pseudo-philosophy of QED, including the Copenhagen Interpretation, has been annihilated. Quantum physics is not the math or the probabilities only, it is not beyond a mechanical interpretation, and it is not fundamentally mysterious. Feynman was wrong: Nature does not refuse to make sense—she is not capricious or willfully irrational. But she does refuse to reveal her secrets, except to those who pay her the proper homage and courtesies. She speaks only to lovers.

EXPLAINING the ELLIPSE

All experiments and observations have confirmed that Kepler's equations are correct and that the shape of the orbit is indeed an ellipse, as he told us. Most physicists have been content to leave it at that. If you are an engineer and you have equations and a diagram, you have all you really need. If you are a physics teacher and you have equations and a diagram, you are well prepared: you can answer almost any question that is likely to come up. But in my chapter on Celestial Mechanics, I showed that the accelerations and velocities in the elliptical orbit were impossible to explain with the gravitational field. That is to say, we have the correct equations, the correct shape, but the wrong mechanics. We have left the equations and the diagram with no foundation for almost four centuries! The proposed and accepted kinematics and dynamics, studied closely, cannot support the motions in the field. Since physics is supposed to be a mechanical explanation of natural phenomena, we have a very real problem here. We have titled this part of physics "celestial mechanics", but we have left out the mechanics almost entirely. This should be a concern to all real scientists, and not just theorists or philosophers, either. If your field does not explain your equations or your diagrams, you are not lacking in metaphysics, you are lacking in physics. What we currently have is a set of equations hanging from sky hooks. A set of free-floating equations is not physics, it is heuristics.

All orbits, whether elliptical or circular, are assumed by historical and current theory to be composed of only two motions, a centripetal acceleration caused by gravity, and a velocity due to the orbiter's "innate motion." This term "innate motion" was most famously used by Newton, and it has never been updated. It is still considered to be the velocity that the orbiter carried into the orbit from prior forces or interactions. It may also be a motion caused by the formation of a nebula or solar disc, but it cannot be caused by the gravitational field of the current orbit. Why? Because there is no mechanism to impart tangential velocity by a gravitational field. Both Newton and Einstein agreed on this. Einstein's tensor calculus shows unambiguously that there is no force at a perpendicular to the field, and Einstein stated it in plain words. How could there be? The force field is generated from the center of the field, and there is no possible way to generate a perpendicular force from the center of a spherical or elliptical gravitational field.

The orbital velocity of an orbiter at any point in the orbit is the vector addition of the two independent motions; that is to say, the centripetal acceleration at that point in the field and the perpendicular velocity, which is a constant. If you study the diagram below, you will find that this can be shown quite simply. The orbiter must retain its innate motion throughout the orbit, no matter the shape of the orbit. If it did not, then its innate motion would dissipate. If it dissipated, the orbit would not be stable. Therefore, the orbiter always retains its innate motion over each and every differential. If we take the two most important differentials, those at perihelion and aphelion, and compare them, we find something astonishing. The tangential velocities due to innate motion are equal, meaning that the velocity tangent to the ellipse is the same in both places. But the accelerations are vastly different, due to the gravitational field. And yet the ellipse shows the same curvature at both places. The ellipse is a symmetrical shape, just like the circle.

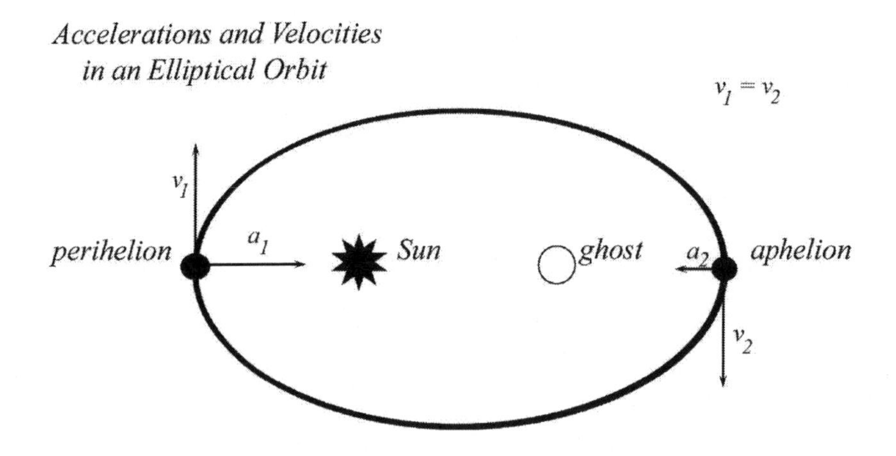

This is physically impossible. Using the given motions, the is impossible to explain. The logical creation of an ellipse requires forces from both foci, but one of our foci is empty. It is a ghost. Every explanation I have seen of the elliptical orbit, including—perhaps most famously—Feynman's explanation, uses the visualization of string and thumbtacks (see diagram above, below title). But this visualization requires two foci. It cannot work with an ellipse and only one focus.

I know that many will cringe that I have claimed in my illustration that $v_1 = v_2$. Don't I know that the orbital velocity varies in an elliptical orbit? Yes I do. Once more, my velocities are not orbital velocities, they are tangential velocities. I refer the readers who do not comprehend my point to my paper on circular motion. In a nutshell, the orbital velocity describes an arc or curved line. It is the vector addition of the tangential velocity and the centripetal acceleration, over the same interval. Newton first created this analysis, and I do not disagree with it. Unfortunately, contemporary physics has forgotten his distinction. It usually conflates orbital velocity and tangential velocity. But the tangential velocity does not curve. It is a straight-line vector with its tail at the tangent. It does not curve even at the limit. It only gets very small at the limit. By going to the limit or to Newton's ultimate interval we do not curve the tangential velocity, we straighten out the arc. That is to say, we straighten out the orbital velocity so that we can apply a vector addition to it, putting it in the same equation as the straight tangential velocity.

Am I saying that celestial bodies cannot be in elliptical orbits? No. I am saying that these elliptical orbits cannot be explained with the theory we currently have. What we currently have is a very complex set of equations for determining the orbits we actually see. This is called heuristics. The theory underlying this math, which is called the theory of the gravitational field, cannot explain the most basic math it contains. From the time of Newton and Kepler, the foundational theory of ellipses has existed with a ghost in it. That is to say, a huge theoretical hole. It is time to fill that hole.

Current theory attempts to plaster up that hole by summing the closed circuit, whether it is circular or elliptical, showing that everything resolves. But this proves nothing, since they cannot help but resolve. We are talking about a closed circuit, by definition. It would be very surprising if the sums did not resolve. What I am talking about here is differentials. Just like in orbital theory, the differentials betray huge holes in the theory. These differentials can be summed, to show a circuit, but the variance they contain cannot be explained by the gravitational field or the innate motion.

To make the ellipse work, you have to vary not only the orbital velocity, but also the tangential velocity. To get the correct shape and curvature to the orbit, you have to vary the object's innate motion. But the object's innate motion cannot vary. The object is not self-propelled. It cannot cause forces upon itself, for the convenience of theorists or diagrams. Celestial bodies have one innate motion, and only one, and it cannot vary.

In this diagram you can see that the vectors given by Newton and Kepler demand more curvature at perihelion than aphelion. When the orbiter is nearer the Sun, its orbital path must show more curvature. The vector v is a constant, by definition or axiom, so the variance in a must determine the curvature of the path at any point. I have also diagrammed the orbital velocities, vo, to show how they are found by adding the other two vectors.

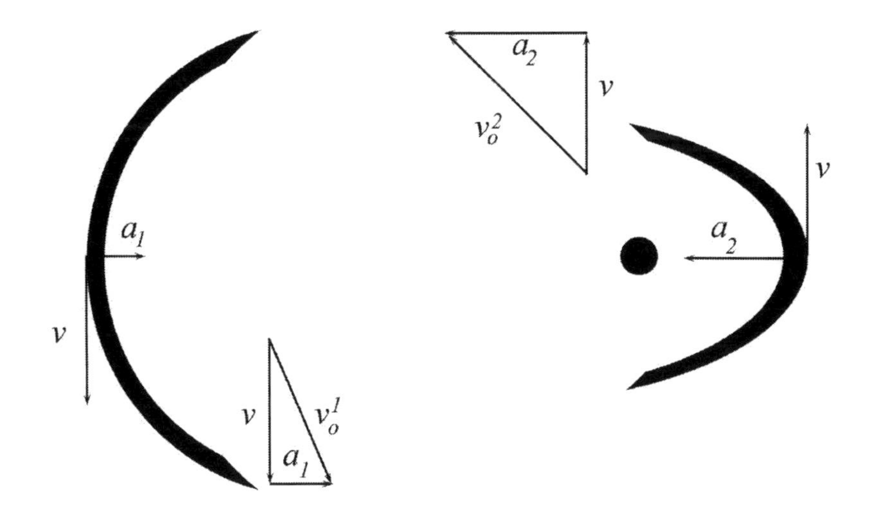

As you can see, the orbital velocity at perihelion is indeed greater than at aphelion, as shown by the length of that vector. But the tangential or perpendicular velocities at all points on the orbital path must be the same. Therefore, we must find the curvatures as I have drawn them here. Now, perhaps, you can more clearly see that these two "ends" of the ellipse cannot be made to meet up. You cannot have greater curvature at perihelion and lesser curvature at aphelion and draw any shape that will meet up. This is my central thesis in this paper. I am not claiming that Kepler's or Newton's math is wrong. I am not claiming that planets do not draw ellipses. Empirically we know that both the equations and the orbital shapes are correct. The problem is with the underlying mechanics. The gravitational field, as it is currently defined, cannot support the shape or the equations. Since the shape and the equations are known to be correct from experiment, we must create a unified field that explains them. I do that in the next section and in my Unified Field paper.

The Solution

Fortunately, the solution is just as simple as the problem. It has been overlooked for centuries, but that does not mean it must be esoteric. It only means that the problem was hidden for a long time. Newton hid the problem so cunningly that no one has detected it since his time.

The solution is that the orbital field is a two-force field. It is not just determined by gravity. Therefore any orbiter must be exhibiting at least three basic motions. The two above, and one other. This other is a motion due to the combined E/M fields of the orbiter and the object orbited. In this case, the Sun and the Earth. The force created by the E/M fields is a repulsive force, like that between two protons. It is therefore a negative vector compared to the gravitational field, which is an attractive field. And so the total field described by gravity and E/M is a differential of the two. In the end, you subtract the E/M acceleration from the acceleration due to gravity.

This explains the ellipse because the E/M repulsive force increases as the objects get nearer. As the gravitational acceleration gets bigger, so does the repulsive acceleration due to E/M.

We have a balancing of forces. This not only explains the varying shape of the orbit, from circle to ellipse to parabola, it explains the correctability of the orbit. It explains why we don't often find orbiters crashing into primaries. It explains how we had a ghost in the other focus of the ellipse: the ghost was inhabited by the E/M field.

This also explains the cause of the ellipse. It has never been understood why some orbits were elliptical and some were nearly circular. Various explanations have been offered, from initial spin, to various perturbations, to an initial angle at intersection to the field. My theory would explain the ellipse in the orbit of captured orbiters by simply showing that the orbiter intersected the field too far from its center. The captured orbiter does not have to intersect the field at just the right distance. It can be captured over a large range of distances, since if it is captured too far away, it will just be thrown into ellipse.

This makes my analysis the opposite of the current analysis. I showed in my Celestial Mechanics paper that current analysis explains the circular orbit as the orbiter intersecting the field at a distance where the two motions balance. By this theory, the ellipse would have to be caused by an initial intersecting radius that was smaller than this balancing radius. I have a diagram in that paper that proves this. If the orbiter is captured at aphelion, for instance, it would begin to get closer to the Sun due to the shape of the ellipse. This could only be explained by showing that the centripetal acceleration overpowered the tangential velocity.

But my orbit is the balancing of three motions, not two. Therefore, the circular orbit would be caused by an intersecting radius where the gravitational and E/M fields balanced. So that to create the ellipse, you would go farther away, not closer. Remember that the E/M field drops off faster than the gravity field. Gravity decreases as $1/R^2$. E/M decreases as $1/R^4$. If you go farther out, gravity overpowers E/M and the orbiter immediately begins to move closer to the Sun.

To show this, I will gloss the capture for an elliptical orbit: 1) the orbiter intersects the field too far away for a circular orbit—meaning that it is beyond the balancing of the three independent motions, but traveling slow enough that the acceleration due to gravity captures it; 2) since the centripetal acceleration initially overpowers the E/M field and the tangential velocity, the orbiter begins to circle closer to the center; 3) but as it does so, the E/M field increases, keeping the orbiter from crashing; 4) the orbiter reaches a minimum orbital distance where the E/M field and the gravitational field [almost] balance; 5) since the orbiter in question is a very large body and the E/M field is made up of very small bodies, the momentum of the orbiter will actually have taken it a small distance inside the balancing radius; 6) the object being slightly below its radius where the two forces balance, the E/M field force is, for a short time, greater than the gravitational force; 7) this creates a very small slingshot effect; 8) due to this effect, the orbiter's momentum carries it outside the balancing radius; 9) if the initial intersection angle was not too steep—so that we didn't get too far under the balancing radius—then we are back to 1). Otherwise we create a parabola instead of an ellipse, and the object escapes a semi-stable orbit. The only step that needs further comment, I think, is step 5. Another way to state step 5 is that the E/M field is a physical object that is much more fluid than the planet that intersects it. The planet is a solid object whose own E/M field is quite rigid. But the central E/M field contains more space and less structure, so that its effect on a solid object will be delayed in this instance.

A useful visualization is to compare the planet intersecting the E/M field to a heavy wooden ball being thrown into deep water. Because the ball is wood, we know that the water will float it—that is, repel it. But if you give the ball enough initial velocity, it will dive into the water to a certain depth before the water begins to reject it. A planet is like a very heavy wooden ball, and the E/M field is like a very weak water. The planet therefore dives to a great depth before the E/M field overcomes the initial momentum. The planet may be "under water" for months. But at last the E/M field floats it.

The buoyancy of the wooden ball determines it force of rejection by the water, and the E/M field of the planet determines its force of rejection by the central field. Its E/M field is determined by its mass and its density.

The visualization is analogous in another way. When the water finally rejects the wooden ball, the ball pops out of the water, often to a measurable height. You have probably experienced this at the swimming pool. If you hold a plastic, air-filled ball under water and then let it go, it will explode out of the water and jump a foot or more into the air. The E/M field of the Sun ultimately rejects the planet in the same way. This is the slingshot effect.

Current theory makes use of this same slingshot effect, but it does not explain the foundational mechanics of it. Current theory tries to build the same unbalanced field as I have, so that the orbiter goes into a sort of gravitational "well." But this unbalance cannot be created with a single field. Any close analysis explodes the whole theory. Current theory has the right effects and the right ideas, it just has the wrong forces. The gravitational field by itself cannot create the forces required to display the effects and curvatures and differentials that are required. To create unbalanced forces and slingshot effects and correctable orbits, you have to have two major intersecting fields. The innate motion is not a field. It is just a simple velocity. In this way it is a constant. It cannot create all the effects that current theory wants to give to the orbit.

Implications

The greatest implication of all this is that Newton's fundamental gravitational equation must be reconsidered. The force in the equation $F = GMm/r^2$ can no longer be considered the expression of a single field. The equation still works, but F must now be understood as the differential between the gravitational field and the E/M field. It is a compound field. All the accelerations we measure are the result of both fields working simultaneously to yield a total force and a total acceleration. This total acceleration is a vector addition of the two constituent accelerations.

A smaller implication is that comets might now be shown to burn not simply from solar radiation, but from the E/M field. That is, the tails of the comet would be produced mainly by electrical considerations. The comet is on electrical fire. This may seem at first to be splitting hairs, but it is not. Solar radiation is not thought to be radiation from an E/M field. It is thought to be ions created as by-products of nuclear fusion. But E/M fields are created independently of nuclear fusion. The Sun would have a powerful E/M field even if it were not a giant nuclear reactor. Therefore, it may be the E/M field that is the main cause for the spectacular effects of comets.

The UNIVERSAL GRAVITATIONAL CONSTANT

It occurred to me some time ago that the Universal Gravitational Constant G might be the key to unlocking the secret to gravity, among other things. It has always seemed puzzling that a constant should have so many unexplained dimensions. A complex constant like this is normally a sign of incomplete theory. All the known concepts are assigned variables and the unknowns are lumped together in a constant. The numerical value is not such a puzzle, since it may just be an expression of incommensurate initial definitions. For instance, we chose the length of the meter and the second and so on pretty much arbitrarily, so it shouldn't be a surprise when all our numbers don't match up at first. But G is not just a number. It has lots of dimensions, L^3/MT^2. Could there be a secret locked up in those dimensions?

I am not the first to ask that question, but no one has yet presented us with any major secrets. Historically, the door for serious questioning was not open that long. Newton's theory became dogma so quickly that very few scientists had the gumption to look hard at it. The ones who did found it mostly convincing or mostly opaque. Since the time of Einstein, no one has taken the constant seriously. It is a piece of discarded and superseded math. To the contemporary physicist, G is about as interesting as the constants of Archimedes or Toltec hieroglyphics. Einstein gave us a new math to express the gravitational field, leaving the mysteries of Newton behind. But Einstein's new math and theory did not dispense with the old mysteries. In many ways it simply changed the text of the mystery. It substituted a new problem for an old one.

I am not concerned with critiquing the math of General Relativity here. By returning to Newton's equations I am in no way questioning the truth or usefulness of Relativity as a whole. I feel I must preface every one of my gravitational and relativistic papers by saying that I am convinced beyond any doubt of time dilation, length contraction and mass increase; I will say it again here. I am returning to Newton's gravitational math not to argue for its historical superiority, but only to answer questions that have remained even after Einstein.

To begin our inquiries, I find it is best to start with a new thought problem. I have solved many other old problems with new thought problems, and I will do so again here. In Newton's original problem there were several unclear points in the definitions or postulates. One of these was whether the distance r included the radius of the large mass. The variable r is supposed to be the distance between two masses, but if the larger mass is very large, its radius comes into question. Another unclear point concerned the smaller mass. In many calculations it is ignored because it is insignificant compared to the larger mass; but you cannot allow it to be zero, for obvious reasons. To avoid getting into the historical discussion of these points, I will offer a thought problem that gets around them completely. In doing so, the thought problem will also bring other things to light.

Let there be two equal spheres of radius r touching at a point. We know that according to the theories of Newton and Einstein there must be a gravitational force at that point, but neither math allows us to calculate it. Newton's math cannot apply since there is no distance between the objects; Einstein's math cannot apply because there is no field at a point. Both theories solve this problem in their own ways, it is true. They add further theory that allows them to calculate in this predicament. In a nutshell they both propose a field centered about a point or a singularity. This causes further problems due to the fact that the objects' gravitational strengths are determined by their masses, and all mass cannot be found at a point. By current theory, mass resides in matter, and matter is made up of atoms. These atoms have real positions: they are found throughout the object—at its outer shell just as at its core. If the mass is a summation of atomic masses, then the force must be a summation of atomic forces. It is difficult to see how the center of force can be behind (in a directional sense) half the mass that causes the force.

We can bypass these further theoretical questions by continuing to propose simple new theory. To do this, let us move our twin spheres s distance apart for a moment. If there is a gravitational force, then after a time interval Δt, this distance will diminish by Δs. Why has the distance diminished? Because a force between the two spheres pulled them closer—this is the classical and current interpretation given to the situation. But can we give it another interpretation? Yes, we can say that both spheres are expanding and that they moved into the distance between them. By the classical interpretation, the centers of the spheres moved toward each other. By my interpretation, they did not.

With my change in theory, you can see that we no longer have to assign Δs to the diminishing distance between the spheres. We can assign it to a change in the radii of the spheres. This being so, we can move the spheres back together, touching at a point. After a time Δt, the radius of each sphere will have changed $\Delta s/2$.

We have changed the idea of gravitational distance in our theory; now let us look at the idea of mass. In article 5 [chapter 1] of Maxwell's *Treatise on Electricity and Magnetism*, he tells us that mass may be expressed in terms of length and time, in this way: $M = L^3/T^2$. He derives these dimensions from a simple substitution into two classical equations.

$$a = m/r^2$$
$$s = at^2/2$$
$$m = 2r^2s/t^2$$

Notice that L^3/T^2 may be thought of as the acceleration of a volume, or a three-dimensional acceleration. This is very suggestive.

This passing idea of Maxwell caused me to reconsider the concept of mass. His math is true, except for one thing. His first equation is not really correct. As written it should be a proportionality. To be an equation requires the constant G.

$$a = Gm/r^2$$

The dimensions of G are L^3/MT^2, which gives the mass and acceleration the correct current dimensions. But what if G is a sort of mirage or misdirection? To pursue this further, I went to Newton's gravity equation, like Maxwell had.

$F = Gmm/r^2$
$ma = 2Gmm/r^2$
$a = 2Gm/r^2$

We must have a 2 on the right side, since the force equation for gravity is the force between two masses, but the force that causes an acceleration on the other side of the equality applies to only one of the masses. It is customary to give all the acceleration to one of the masses, but in my thought problem the two equal spheres both accelerate. Now let us apply this equation to our twin spheres touching at a point. There is no distance between the spheres, so r would normally apply to the distance from center to center. But since the spheres are the same size, let us re-assign r to the radius of each sphere. The distance from center to center is then 2r. We have assigned Δs to a change in the radius instead of a change in the distance between the spheres, and this allows us to calculate even when the spheres are touching. For clarity let us make Δs into Δr.

$a/2 = Gm/(2r)^2$
$a/2 = \Delta r/\Delta t^2$
$\Delta r/\Delta t^2 = Gm/r^2$

The only remaining problem is the variable r. If the spheres are expanding, then r must be expanding. After time Δt, the radius will be $r + \Delta r$. After any appreciable amount of time, r will be negligible in relation to Δr, so that $\Delta r \approx r + \Delta r$. Therefore we may simply drop the r variable as a variable that approaches zero.

$\Delta r^2/\Delta t^2 = Gm/\Delta r$
$m = \Delta r^3/G\Delta t^2$

Now all we have to do is reassign the dimensions of L^3/T^2 to the mass, as Maxwell implicitly suggested. We will drop the dimension M altogether. This gives G no dimensions at all. It is just a number. This is actually much more sensible, since constants with dimensions are a sign of incomplete theory. That is what drew me to this solution in the first place. Newton had to give the dimensions L^3/MT^2 to G only because he had mistakenly assigned mass a new dimension. Mass is not a new dimension. It is reducible to the old fundamental dimensions of length and time.

Our last problem is plugging known values into this new equation. At first it looks like the mass should be changing over time, since the radius is changing. But no. The mass is dependent on $\Delta r/\Delta t^2$, and that is not changing over time. As the radius gets larger, so does the change in time, so that the ratio is constant. It is a constant acceleration. A constant acceleration gives us a constant mass. Therefore we can plug known values for m into this equation.

$m = \Delta r^3/G\Delta t^2$
$a = 2\Delta r/\Delta t^2$
$a = 2mG/\Delta r^2$

That is the acceleration of each of two equal masses in a gravitational situation. But if we want to give all the acceleration to one of them, holding the other one steady for experimental purposes, then we simply double the value.

$a = 4mG / \Delta r^2$

If the proton has a radius of 3.17×10^{-13}m*, this yields

$a = 4.44 \times 10^{-12} m/s^2$

We are now in a position to use our new number for acceleration to explain a current experimental mystery. Using the new number to do this will also act as proof of my theory, since it gives us a sort of experimental confirmation.

If the proton has a fundamental spherical acceleration,* then in any one direction it will have a velocity at any given time. If we suppose that the age of the proton is on the order of the age of the universe, then we can estimate the current velocity of the shell of the proton. "Velocity relative to what?" you may ask. "If everything is expanding, then what is our background?" The velocity we will find must be relative to two things. It is relative to the velocity of the radius at t_0, which we define as zero. And, it is relative to the speed of light, c. Einstein defined the speed of light as the universal background, and I continue to accept that definition.

If we accept (one of) the current estimates for the age of the universe as around 15 billion years, then the current velocity of the proton's shell would be 2.1×10^6m/s.

$$v = at = (4.44 \times 10^{-12} m/s^2)(4.73 \times 10^{17} s) = 2.1 \times 10^6 m/s$$

That seems ridiculously large at first, except that we have experimental confirmation of that number from accelerators. As I have shown in my paper on accelerators, there is a limit to the speed achieved by the proton. This limit is a final energy of about 108 times the rest energy. Using *gamma*, this translates to a velocity of .999957c, which is 1.2×10^4m/s short of c. If we theorize that the gap between c and the limit in velocity is caused by a residual velocity or velocity equivalent that the proton already has, then the limit is explained.

But there is more. My correction to *gamma* and to the mass increase equations predicts a limit in velocity for the proton of .9930474c, which is 2.1×10^6m/s short of c. This is an exact match, as you see. If we plug the current form of *gamma* into my acceleration equation above, we get an age of the universe of only 85 million years. But my correction to gamma gives us an age of 15 billion years. We know that protons must be older than 85 million years. The Earth is almost 60 times older than that itself.

Now let me address the first outcries. Some will say that my equations above give us a current radius for the proton of .7 meters. But that is bad math. You have to do a proportionality. As much as .7m is larger than x, x is larger than the proton originally was. If we say that the current proton radius is 10^{-13}m then the original proton radius would have been

$.7/x = x/r$
$r = 10^{-26}$m

The next problem concerns my claim, just made, that velocity is constant. A velocity, such as the speed of light, will remain constant in an expanding universe simply because time is a function of distance. What I mean is that we define time in relation to distance. If this definitional distance increases, as everything expands, then the definitional period of time will increase proportionally. Distance gets larger, time gets "larger".

So the ratio of the two stays the same. Which means that all relative velocities will stay the same.

As an example, we now use the cesium atom to define time. The baseline data in the cesium atom is an oscillation from one energy level to the other, or an atomic wobble. This oscillation is a motion, and all motion implies a distance. If the cesium atom gets bigger, then the distance increases, and the time period increases. Time is dependent on distance. This is even clearer with a pendulum clock. If all material lengths increase, then the length of the pendulum will of course increase, which will increase the length of the second. Time is connected definitionally and operationally to distance, therefore any increase in universal length will cause a proportional increase in universal time. Since velocity is defined as one over the other, velocity will not change. The numerator and denominator both get bigger at the same rate.

* See my paper on the Bohr Magneton.

CELESTIAL MECHANICS
unanswered questions

Celestial mechanics has not made much progress since Kepler and Newton. Even General Relativity only recast the old concepts in new but basically equivalent terms. Einstein did not overthrow the fundamental mathematics of gravity and orbits. The old conceptualizations and equations still stand; they are still taught in schools everywhere. General Relativity only fine tunes them, by substituting a different but basically equivalent theory (curved space for action at a distance) and a nearly equivalent mathematics (tensor calculus for calculus). Einstein never implies that Kepler and Newton's theories were wrong—they are only incomplete. Import the finite speed of light and the tensor calculus into classical theory and you have current wisdom with regard to celestial mechanics.

Kepler's laws still hold, Newton's laws still hold. General Relativity and contemporary celestial mechanics take them as givens, as starting points. For instance, Kepler's theory of ellipses still pertains to this day. One might say it remains the bedrock of contemporary celestial mechanics. Richard Feynman recalculated Newton's proof of the elliptical orbit using only plane geometry in his famous "lost" lecture. He had nothing to add but an updated proof. And General Relativity never questions accepted concepts like the theory of ellipses. For Einstein, the gravitational field remains a Keplerian beast, in shape and size and influence. The only difference is in calculating specific accelerations within that field

There is one further difference of course: the genesis of that field. Kepler and Newton believed that a gravitational field was produced by a massive object, that space (if not the field) was rectilinear, and that the massive object acted directly—though in an unknown way—upon any matter within the field. Einstein changed all that, though in a less drastic way than is commonly assumed. He agreed that the field was produced by the massive object, but he theorized that the object acted on the field rather than on matter in the field. This produced a spherical field, which then acted on any matter within it. He went even further, though, for he believed that "the field" and "space" were two words for the same thing. For him they were equivalent abstractions or ideas. If the field around a massive object was curved, then space was. There was nothing left over, nothing that you could call space after you defined the field.

Notice, however, that the gravitational mechanism remains equally mysterious. Newton could not say how a massive object acted upon matter at a distance. Einstein cannot explain how a massive object curves space at a distance. There is much talk and work currently on gravitons, but none have been found. And Einstein never presented them as the mechanism for gravity anyway. He postulated that gravity waves might be produced under certain situations, and that the waves might be composed of gravitons, but he never imn beast, in shape and size and influence. The only difference is in calculating specific accelerations within that field.

There is one further difference of course: the genesis of that field. Kepler and Newton believed that a gravitational field was produced by a massive object, that space (if not the field) was rectilinear, and that the massive object acted directly—though in an unknown way—upon any matter within the field. Einstein changed all that, though in a less drastic way than is commonly assumed. He agreed that the field was produced by the massive object, but he theorized that the object acted on the field rather than on matter in the field. This produced a spherical field, which then acted on any matter within it. He went even further, though, for he believed that "the field" and "space" were two words for the same thing. For him they were equivalent abstractions or ideas. If the field around a massive object was curved, then space was. There was nothing left over, nothing that you could call space after you defined the field.

Notice, however, that the gravitational mechanism remains equally mysterious. Newton could not say how a massive object acted upon matter at a distance. Einstein cannot explain how a massive object curves space at a distance. There is much talk and work currently on gravitons, but none have been found. And Einstein never presented them as the mechanism for gravity anyway. He postulated that gravity waves might be produced under certain situations, and that the waves might be composed of gravitons, but he never implied that a normal gravitational field was produced by gravitons.

In saying that a massive object curves space, Einstein was in many ways begging the question. He was removing the problem one more step. For Newton, the mystery was in understanding how the Sun influenced the Earth, for instance. For Einstein, the mystery becomes in understanding how the Sun influences the space around it, which then influences the Earth. It is a sort of *obscurum per obscurius*—explaining the obscure by use of the more obscure. We are now taught, in courses influenced by the thinking of Einstein, that the elegance of a scientific theory resides, in part, in its simplicity. Given two theories that have the same content—the same power of prediction—always choose the one that has the fewest moving parts, the fewest postulates. General Relativity fails on this basis alone; it is cut by Occam's Razor. It not only fails to solve the problem of Kepler and Newton, it adds to it. The mystery of influence remains unsolved, and the mechanism now has two steps rather than one.

What I will show is that Kepler and Newton, although mathematically correct in most basic ways, left us with underlying theory that was incomplete. Einstein perfected the math, but left the underlying theory almost untouched. Relativity gave him the tools to fill in the conceptual holes of classical gravitational theory, but he did not use these tools to their full effect. Diverted by the tensor calculus, he lost sight of some of the simple conceptual shortcomings that his theory should, and could, have addressed. This has left us with orbital math that is a very precise heuristics. That is, it allows us to express empirical data with great accuracy. But it does not show why the empirical data is what it is. Its failures are the same failures as classical theory. [GR has some mathematical failures, too—the most important of which is the failure of *gamma*—but those are addressed in other papers.]

Ultimately I must take exception to Kepler's theory of ellipses [chapter 12]. But to do this, I must go back even further. I must start with a single object orbiting a central mass, an Earth orbiting a Sun in a perfect circle, such as Archimedes might have understood. In this ultimately simple version of an orbit, we have only two velocities. We have a tangential velocity and a centripetal acceleration—which causes a so-called instantaneous centripetal velocity. Newton assigned the centripetal acceleration to gravity and the tangential velocity to the orbiting body itself. That is, the tangential velocity is not caused by the gravitational field. How could it be? It is perpendicular to that field, whether the field is rectilinear or curved. It is stated explicitly that the Earth had this velocity before it entered the orbit. Newton calls it the body's "innate motion." A gravitational field has no braking effect; therefore, since a body retains a velocity until another force acts on it, the Earth still has the velocity in orbit. Notice that if the Earth had no velocity tangential to the Sun's gravitational field as it was captured by that field, it would simply crash directly into the Sun. So the Earth must have an initial tangential velocity, and it retains this velocity after it is captured by the Sun. This velocity is the velocity shown in all current and historical illustrations, one and the same.

As I said above, this analysis began with Newton when he described circular motion in Proposition I of *The Principia*. The orbiting body is assumed by Newton to have a velocity due to "its innate force." So this motion must be independent of the gravitational field. His assumption has never been seriously questioned.

When we are shown the illustration of circular motion in our physics textbooks, we are always shown the accompanying illustration, which is that of a ball on a string. The boy whirls the ball around him, and a circular orbit is created. The force that the boy's hand must exert on the string is analogous to the gravity of the Sun, we are told. The swinging action of the boy creates the tangential velocity. So in this case, the hand creates both velocities. In fact, there is a dependence between the tangential velocity and the centripetal acceleration, a dependence given mathematical form by the equation $a = v^2/r$. But in the illustration of the orbiting Earth, the Sun does not swing the Earth—there is no implication of that. The tangential velocity and the centripetal acceleration are completely independent. There is no string or other force that could impart tangential velocity to the Earth. Assuredly, the Sun is spinning, and this may create tangential perturbations in an accompanying E/M field; but there is no way, in this simplified illustration, that the Sun could be the cause of the tangential velocity of the Earth. And if the Sun is creating tangential perturbations in the gravitational field, the theory must mechanically explain how they are produced. **No theories have ever done this.**

Even greater problems arise when we try to imagine how the Earth was captured by the Sun. How is an orbit like this created? How is any planetary orbit created? The textbooks never go there. By giving us the ball-on-a-string illustration, the book leaves the impression that the analogy is complete; that is, that the tangential velocity and the acceleration are conceptually connected in both instances. We are left with a *fait accompli*: since the two motions are tied to one another with the ball on a string, the two motions must be tied in the Earth/Sun example, and there is nothing to explain. But there is an awful lot to explain. To start with, in reality an orbit like this creates a hairline balance of two independent motions. The tangential motion and the centripetal motion must be perfectly balanced or the orbit will deteriorate immediately in one direction or another (inward or outward). Any satellite engineer knows this. There is one perfect distance that creates a stable orbit for a given velocity. Any other orbit requires the satellite to speed up or slow down—to make corrections. Obviously, the Earth cannot make any corrections. It is not self-propelled. It cannot speed up or slow down. Therefore it must be taken to its optimum distance and kept there.

Now, think of the Earth's orbit for a moment. Let's work backwards and see if we can imagine how the Earth might get to that optimum distance, with just the optimum tangential velocity. If you reverse time, and conceptually back the Earth out of orbit, you see that the only way you can do so is if you accelerate it out of there. If you keep the same velocity, it stays in orbit. If you decelerate, then it crashes into the Sun. So you must accelerate the Earth out of the orbit. But that means that unless the Earth was ejected by the Sun, it had to decelerate to reach its present position. If it is coming from outer space into the field of the Sun, it must somehow decelerate in order to fall into its current position. But how can an object entering a gravitational field decelerate? It is getting closer to the Sun: it should be accelerating. The only possibility appears to be a fortunate collision that accidentally throws it into the perfect spot. Even a planet ejected by the Sun cannot reach any possible orbit, without a collision, since an ejected planet will not have any velocity tangential to the Sun. There is no way to eject an object from the center of its future orbit with a velocity tangential to that orbit.

So, the unavoidable implication of historical theory is that all orbits must have been created by fortuitous collisions, either by planets arriving from outer space or being ejected by the Sun. The problem is that planets arriving in orbits immediately after collisions are going to be damaged planets. Most likely they are going to be out of round. They are going to be missing chunks. This is a problem since imperfect planets create perturbations in orbits. Spins and wobbles are created, which cause uneven velocities and uneven forces. This should be fatal since the sort of orbit described by current theory is not correctable. There is no margin of error. Either the forces balance or they do not. If they do not, then the orbit should not be stable.

Some will interrupt here to point out that current theory provides that the Earth was formed from a solar disc. It was not captured or ejected; it was simply always there, in some form. It congealed out of the nebula. But this answers nothing, for current theory fails to explain how this primordial disc of pre-planets or planetoids achieved its tangential motion in the first place (see below). Gravitational theory provides absolutely no mechanism, not even one as magical as gravity, to explain rotational motion in a gravitational field. It is the same question as to why galaxies rotate like wheels: they just do. We have a partial answer for why the stars don't fly out into space: gravity. But we have no answer at all for why the stars move sideways to the gravitational field of the galaxy. If they weren't captured, what set them in motion? The pat answer is "a spinning gravitational field", but if you ask how a gravitational field imparts tangential velocity you get no answer. It is implied that the spin of the Sun about its own axis somehow set the whole solar system to spinning, but this is mystical in the extreme. Almost no one thinks that the Moon's orbit is caused by the rotation of the Earth about its own axis. No one thinks this because there is no mechanism to link the rotation of the Earth to the orbit of the Moon. There is no mechanism to link the orbit of the solar disc to the spin of the Sun either, and yet it is accepted at face value.

All the other perturbations of the solar system are likewise mysterious. The planets affect each other by applying small torques to one another, we are told. How can you postulate the applying of torques with a gravitational field—a field that is absolutely incapable of creating mechanical torques? Current celestial mechanics discovers the perturbations, gives them mathematical form, but does not explain the mechanics of the perturbations. It would be better labeled Celestial Heuristics.

According to current theory, gravity is either an attractive force or a space warp. In neither case can you mechanically explain a torque. The field is generated from its center and cannot possibly do anything but pull inward from that center. Even with a spinning gravitational field, no torque is possible. We are told that angular momentum is carried out to orbiting bodies, but how? It cannot be via the gravitational field. There is no proposed mechanism. Einstein expresses known forces with tensors, but he cannot explain the genesis of those tensors. Where do the tangential components of the tensors come from? We don't get so much as a theory. Nothing. That is the main reason physicists have added the graviton to the fundamental field of gravity, despite the fact that Einstein assured them that objects in curved space "felt no force," and despite the fact that they still parrot this claim—believing that GR is geometric, not force-carrying. They need the graviton to help them explain torques.

The graviton would not help them anyway. A torque could be applied by an exclusionary field—like the E/M field. But a torque could not be applied by an attractive or warping field. The graviton, if it existed, would cause attraction or the equivalent of attraction. Even if the graviton carried angular momentum, it wouldbe a sort of negative angular momentum, like the negative force it carried that caused the body to come closer. This would put all objects in retrograde orbits, and we don't see this. We don't see negative torques, we see positive torques—prograde torques.

Mean motion resonances are also impossible to explain with gravitational fields, for the same reason. Gravity is a centripetal force, not a tangential force, so that resonances are beyond explanation. This also applies to tides and equatorial bulges. No one doubts they exist, but how can gravity explain them? How can curved space explain tides? Beyond curved space you are back to force at a distance—not only centripetal force but tangential force. You must have angular momentums working at a distance. How?

Another problem is that even the current model believes that some satellites, like Triton and Phoebe, are captured satellites. Captured satellites must have been captured as I stated above—by decelerating into orbit. How was this possibly achieved, given the current list of forces and causes of forces? What, exactly, caused Triton to settle into its current orbit? A balancing of instantaneous velocities cannot explain it, since even if Triton happened to intersect its future orbit at exactly the right distance and at a precise 90° angle, many other factors would also be involved. Neither Triton nor Neptune is an ideal body. They both would have had some spin. Just as an example, it is believed that all bodies apply torques to all other bodies (although it is not explained how in current theory). Therefore Neptune must ha ve a rather complex field at all orbits, not just a simple centripetal acceleration. Scientists use this complex field to explain the motions of Neptune's other Moons. If you add this complexity to the real field of Neptune, you see that the odds of Triton arriving with all the perfect counter-speeds and counter-torques, at just the right angle and distance are precisely zero. There must be some correctability to orbits not only to account for the stable orbits we see but also to account for the creation of captured orbits. The field of Neptune must have some ability to resist small deviations and to correct them. Otherwise no body could ever be captured in the first place.

It is true that the orbit of Triton is decaying, so that the orbit is not in fact completely stable. But this is not the question. No field is infinitely forgivable, but orbits show a degree of float that is not in line with current theory. There appear to be constraints on decay and escape far beyond what would be logically expected. A decaying orbit like Triton's would be expected to fail exponentially. As Triton lost energy it would fall into a lower orbit. At this lower orbit the acceleration toward Neptune is even faster. To be in a stable orbit at a smaller radius, Triton would have needed to gain energy, or speed up. It has slowed down and gone lower, therefore we would expect a multiplied affect. Instead we see a long slow decay. Once again, empirical evidence directly contradicts the given theory of gravity and orbit.

A similar problem is caused by any three-body analysis. Insert even one Moon into a planetary orbit that is the balance of a tangential velocity and an independent centripetal acceleration and you have a crash. A Moon creates a perturbation that cannot correct itself. For instance, take the familiar two-body illustration and add a 3rd body. Say this 3rd body is the Moon, and put the Moon between the Earth and the Sun. We know the Moon goes there occasionally because we see total eclipses. Well, the Moon is going to pull the Earth into a fractionally lower orbit. Physicists have never explained how this is not fatal to the orbit. They know that it is not fatal, since the Earth does not crash into the Sun, so they simply do the math to explain how the Earth gets to the next position that it actually achieves. But to do this they must give the Earth slightly eccentric little accelerations and decelerations, which they never explain. They give the Earth a little tug here and there, saying that the Moon corrects for itself. But this is absurd. A balancing of velocities like this cannot be self-correcting.

For instance, if you think that the Moon simply pulls the Earth back out of danger two weeks later when it is farthest from the Sun, you are not thinking right. The Moon has pulled the Earth closer to the Sun: in order for it to now pull it back two weeks later, it would have to be bigger. It takes a greater force to nudge a planet into a higher orbit than it does to nudge it into a lower orbit. And the same problem is going to be met when the Moon is sideways to the Earth. It is going to slow it down and then speed it up two weeks later. All these perturbations cannot be made to offset. No matter which direction you have the Moon going (clockwise or counter) you are going to have the Earth thrown into ever lower orbits for two straight weeks. The next two weeks of corrections cannot offset this. And this is not even taking into account the Sun's effect on the Moon's orbit, which causes further uncorrectable perturbations.

You may say that I am taking only the case where the Moon is orbiting in the plane of the Earth's orbit. But there is no plane of orbit that is self-correcting in this situation. I encourage you to try it.

The usual answer to this is to show a summing of potential and kinetic energies in a closed loop and prove mathematically that all energy is conserved. But this fails to address the issue. I am not complaining here about a sum or an integral. Mathematically I am pointing at differentials. If you look at individual motions in any orbit that has three or more bodies, you will find that the differentials show a variation in the tangential velocity of the orbiting body. But natural bodies like planets and stars and Moons cannot vary their tangential velocities on demand of the math. As I said, they are not self-propelled. They cannot make any corrections. If the differentials are showing a variation, this variation must be explained by an external force. Gravitational theory gives us no force to explain it. Neither Newton nor Kepler nor Einstein have anything to say on the subject. It is one of the great unseen gaps in kinematics.

This is not to say that the math is incorrect. It isn't. It is simply unsupported. We have failed to build an orbit that is correctable or stable. Our engineers can build a stable orbit, our mathematicians can build a stable orbit, but our theory cannot yet do so.

The history of celestial mechanics is a history of mathematical analysis that is very short on theory. Every book you will find in the section on celestial mechanics at even the largest university libraries concerns creating equations to explain orbits based on observations. Three or four observations allow you to build a basic equation. Most books have differential equations on the first page, and those that don't begin by glossing the history from Newton to Gauss—a history of mathematical analysis. Most books don't have a single page on the theory of orbits. That is because no one has done theory since Kepler and Newton. The problems I am enumerating here are mostly not known to exist anymore, for the very reason that all study of orbits and gravity is now strictly mathematical. No one cares "why", they only want to discuss "how". If there are huge holes in the gravitational theories of Newton and Einstein, what does it matter? We have a heuristic theory that allows us to put our own objects into orbit, what else do we need?

There are many other similar mysteries about the stability of orbits, but I think I have made my point in regard to the circular orbit. Let us now graduate from the mysteries of the circular orbit to the mysteries of the elliptical orbit. I have already showed you how to solve the ellipse in the previous chapter, but we may look at further problems with the ellipse here. What we imagine when we accept the ellipse as a logical-looking orbit is that it is simply a sort of squashed circular orbit. We think, well, maybe when a planet is captured, it first hits an orbital tangent at an angle, instead of at a perfect perpendicular. This throws its orbit a bit out of whack, but the orbit is somehow stable since the total area of the orbit is about the same. All very unscientific, but I would guess that many of us have assumed these things, without really questioning it very deeply. But, let's build that ellipse again, starting from aphelion. Let us draw the whole thing, just accepting that an ellipse must somehow be created, since we have evidence of them in the solar system. Finally, let us look for the "equivalent" circular orbit. Meaning that if we have the same planet with the same initial velocity and we want to put it into a circular orbit, where do we put it? Turns out that the circle is completely outside the ellipse, and that it has a lot greater area.

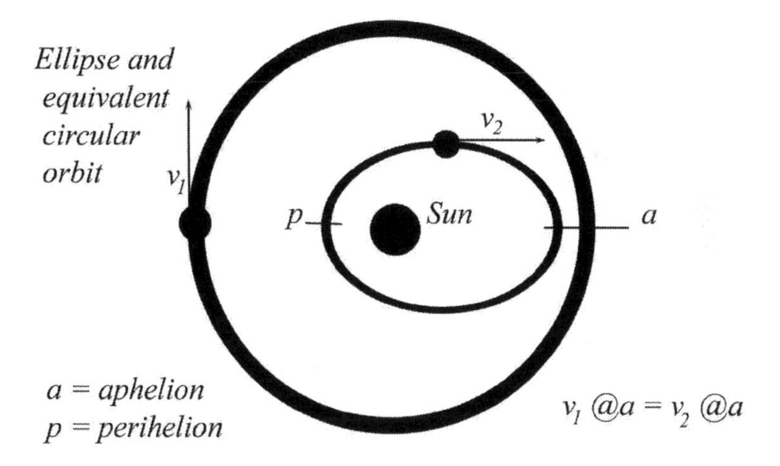

Remember that the only way we can explain the planet in ellipse beginning to dive toward the Sun as we move it past aphelion is that its velocity is not great enough to keep it in circular orbit. Therefore, to put it into a stable circular orbit, we must move it further away from the Sun at aphelion. If we do that then aphelion becomes the radius of the circle, and we have our circular orbit. As you can see from the illustration, the path of the ellipse never crosses the path of the "equivalent" circle. If that is true, then the planet in ellipse can never reach a point where its perpendicular velocity overcomes the centripetal acceleration produced by the gravitational field. It never achieves a temporary escape velocity. No, it simply spirals into the Sun. Its orbital velocity increases, yes. The "orbital velocity" continues to increase until the planet burns up in the Sun's corona.

Here is another argument against current theory. Consider Kepler's Third Law. It states that the ratio of the squares of the periods of all potential orbits are equal to the ratio of the cubes of their average distances. This law is still accepted. Einstein accepted it. It is in all current textbooks. Furthermore, it is confirmed by the most exacting modern measurements. To within a small fraction of error, the ratio r^3/t^2 for the nine planets is $3.34 \times 10^{24} km^3/yr^2$. What this means, of course, is that the orbit of the planet has nothing to do with the mass of the planet. According to Kepler's law, one must balance only the distance and the period. To see what I mean, take the Earth out to the distance of Jupiter and try to build an orbit. Could you do it? Of course. You just slow the Earth's orbital velocity down until it offsets the centripetal force from the Sun. What you find is that the Earth will match the orbital velocity of Jupiter exactly. Somewhat surprising, isn't it? I assume that some readers will have thought that the Earth would be going slower, since it is smaller. It feels a smaller force from the Sun, therefore it has less centripetal acceleration to offset with its velocity. But that is not how a gravitational field works. Yes, the force is different, but the acceleration is the same. F = ma. That is why all objects fall at the same rate in a vacuum, remember? Jupiter and the Earth fall toward the Sun at the same rate—that is, the same acceleration—if they are at the same distance. You will say, "But the Sun must pull harder on Jupiter, surely, to keep it in orbit, than on the Earth." Yes, surely. And that is my point. A gravitational field is a strange creature, and its characteristics have never been explained. They have been *described*, in several different ways, by Newton, Einstein, etc., but never explained. The gravitational field is not a force field, it is an acceleration field. When Newton or Einstein maps the varying numbers at varying distances in the field, he is mapping accelerations, not forces. Very mysterious, that. Notice that acceleration is a measurement of rate of change of motion. Acceleration is not directly a measurement of force. Movement, not force. That is very important.

Nor has this problem been solved by General Relativity. More money is now being spent worldwide on finding the graviton than on any other scientific project. Billions, literally. It will not be found, but a good question to ask those who seek it is this: Would the Sun need to send out bigger or more powerful gravitons to Jupiter than to the Earth, if they were both at the same orbital distance? If so, how does the Sun know which to send?

Perhaps we need to look for a messenger particle, one that precedes the graviton, and asks the orbiting object how much it weighs. I know that this all sounds like a joke, but the question must be addressed seriously by those who put "no action at a distance" on their t-shirts. The status quo in physics, made up of the biggest names in the field in the 20th century, still brags about this in the latest books. But their theories explain absolutely nothing.

You may be asking yourself at this point, how has all this sloppiness stayed buried for so long? Stephen Hawking told us just twenty years ago that we were a decade away from knowing everything. The end of physics. Except for chasing the graviton, no one is even working on gravity anymore. It is a problem that is considered solved. The "great minds" are busy with superstring theory, and things like that. Tying gravity to quantum mechanics. But here I am saying that no theoretical progress has been made since Newton. How can that be? One word. Obstruction.

The obstruction began with Newton himself. Newton derived Kepler's law from his own, to show that the two were consistent. He did it like this, roughly.

Given that $f = ma$ and that $F = Gm_1/r^2$ (Newton's famous equations, of course) Let $f = F$

$Gm_1m_2r^2 = m_1a$
Then let $a = v^2/r$
$Gm_1m_2/r^2 = m_1v^2/r$

Next, since all the orbits of the planets are nearly circular, let the distance traveled in each orbit equal the circumference of the orbit:

$v = 2\pi r/t$

$Gm_1m_2/r^2 = m_1[2\pi r/t]^2/r$
$Gm_2/r^3 = 4\pi^2/t^2$
$t^2/r^3 = 4\pi^2/Gm_2$

Since the right side is a constant for all planets around the Sun, the left side applies to all the planets, and all possible planets.

This derivation is problematic not because we let the orbit be circular. That was only to simplify the math. The problem is in letting $a = v^2/r$. As I showed above, this equation is applicable only when a is dependent upon v. If Newton or current textbooks want to use that equation, then they must explain how a is dependent upon v. Newton is implying that there is a necessary causal connection between the two, without providing us with a means of causation. For, I repeat, how can a gravitational field cause a velocity tangent to that field? Or, to make the analogy even tighter, how can the tangential velocity determine the field strength? That is what is happening with the ball on a string. Increased velocity causes a greater force on the string.

To show this, let us look more closely at the Full Moon position. By the equation $a = v^2/r$, the centripetal acceleration of the Earth is .006016 m/s^2. The centripetal acceleration of the Moon is .006407. The Moon is outside the Earth, but feels more pull from the Sun? That is impossible. The gravity field decreases with increasing distance, remember? That old inverse square law? How and why would the Sun pull more on a body just because that body sped up? The orbital equation is hiding a huge hole in the field mechanics. In fact, Newton's orbital equation contradicts Newton's gravitational equation. Newton's gravitational equation tells us that the field diminishes by the inverse square, but the orbital equation tells us the field can increase at greater distances simply by increasing the speed of the orbiter!

As a final example of the current state of the art in celestial mechanics, let me show you a specific example from *The Encyclopedia of the Solar System*, a recent book [1999] published by NASA and the Jet Propulsion Lab with the cooperation of many of the top universities in the country. To tie the Sun to a solar disc, we are given evidence from T Tauri stars. These stars are of roughly solar mass and they have what appear to be discs. Or, to be more precise, they have a lot dust around them. There is enough dust to obscure the stars, but it doesn't. Why? Because it is confined to a disc and the disc isn't in our plane of sight. So far so good. To confirm this, we look at the emission lines created by a stellar wind. We find that these emission lines are always blue-shifted. And this is where it gets silly. For the book tells us: "This observation is explained if the red-shifted lines that would be associated with gas flowing away from the observer were obscured by a circumstellar disc." Anybody see the problem here? There are actually two problems. The first is that if the disc is not in our plane of sight, then it can't be the cause of any obscuring of shifts, red or blue. The second problem is that the gas flowing away from the observer is on the far side of the star. If light from the star goes through it in order to make any emission lines, then the light must be going in the opposite direction of the Earth. We can't possibly see it. This whole theory is a comedy of basic logical errors.

It is not the exception, either, it is the rule. A mistake like this cannot be assigned to single person. This book was edited by a large committee of top-flight physicists. Besides, contemporary physics is riddled with basic mistakes like this, mistakes that are nothing less than shocking.

If physics is to regain any sort of health, it must begin to take mechanics and conceptual analysis seriously again. And it must regain a degree of rigor and self-criticism. Currently it is awash in a sea of self-glorification. Lists of things that are still unknown are occasionally published, but substantive papers pointing out the very real faults of the Standard Model are dismissed without a reading and their authors are blacklisted. Besides, these lists of unsolved mysteries are always long on big theoretical problems—the sort of things that might be expected to remain even though we are brilliant masters of the universe. They are short on specific holes in existing theory. So you hear a lot about how we can't yet explain the total amount of matter in the universe, but you hear nothing about how we can't explain why natural orbits don't immediately deteriorate or how we can't explain force fields. We are being purposefully misdirected away from the real problems.

THE SUN'S LACK OF ANGULAR MOMENTUM
explained by the unified field

It is admitted by most that current models cannot explain the lack of angular momentum in the Sun. But using my Unified Field, we have a straightforward mechanical explanation. We have seen that the motions of planets are determined by both gravity and the foundational E/M field or charge field. The planets have more angular momentum because they have been gaining it all the time, simply from being in orbit. The E/M field of the Sun gives a constant torque to every planet in the system, and those planets give torques to their satellites. It is precisely this torque that causes and maintains the orbital velocity of the orbiter, as well as its spin. The standard model could never explain the cause of torques, resonances, or spins, and it is precisely because they didn't know that they had the charge field to work with.

Then why don't all orbiters have great angular momentum? The Moon, for example, has almost none [moment of inertia =.391]. It is showing its same face to us all the time. Other satellites are also in this type of orbit. Why is this so? If primaries are always applying torques, then satellites would all be expected to be spinning very fast, right?

Well, no. Here again there is an easy answer, but you have to know something about the particular orbit and the bodies involved. In the case of the Moon, most of the torque from the Earth has gone into velocity rather than spin. The angular momentum of Earth+Moon is very high, and this is not due to the size of the Moon alone. The orbital speed of the Moon is also very great, given its size and the size of the Earth. But why does the Moon resist spinning? Simple again, since the Moon is not evenly weighted. That is, its center of mass is offset appreciably from its center of figure. As would be expected from my theory, this center of mass has positioned itself as close as possible to the Earth. I say my theory, since it is not clear how GR can explain things like this. An orbiting object in a curved trajectory that was feeling no forces could hardly re-center itself in regard to a distant object.

Classical theory could explain it as an unequal attractive force, and this would bring the center of mass toward the Earth. But classical theory could not explain anything beyond that, including any torques, and therefore could not explain why an uncentered mass would resist spin in this situation. GR can explain even less, offering us tensors that express the uneven force but do nothing to explain its genesis. My theory explains it once again as a joint effort of real acceleration from the center of the orbit and torques caused by the E/M field. The Moon also has increased density toward its center, and a relatively low overall density [3.3g/cm^3], both of which make it resistant to a spin-inducing torque in this situation. Over time, the Moon has found it more efficient—for a number of complex reasons that require a close analysis of the intersecting E/M fields, as well as the field of the Sun—to channel the torque into orbital velocity rather than spin. In a nutshell, several mechanical and physical factors resist spin and these same factors do not resist velocity. This is proved by the Moon's increasing orbit: the torque is pushing the Moon into a higher orbit rather than giving the Moon more prograde spin.

There are different factors limiting the spin of the Earth. The Earth is better balanced than the Moon, and therefore it spins quite quickly. It would spin even more quickly, due to a strong torque from the Sun, if it did not dissipate much of its angular momentum into the Moon's orbital velocity, via the torques we have been talking about. If the Earth had all the angular momentum that the Earth+Moon now has, it would spin once every 4 hours. So you can see that much depends on the structure of the bodies in question. But in general, the planets with their satellites have much more angular momentum than the Sun simply because the Sun is constantly applying a torque to them.

Current theory provides many dissipative forces to explain the loss of angular momentum in the Sun, including non-magnetic turbulent friction, magnetic coupling, and other desperate ad hoc theories. But the fact is that no dissipation is necessary if the Sun is never assumed to have had the high angular momentum that the planets now have. Current theory simply made the wrong first assumption. It assumed that the Sun must have originally had the same angular momentum as the planets. I have shown that this is upside down. The planets, if formed by the nebula and disc, would have originally had the same low angular momentum as the Sun. But they have gained momentum and the Sun has not.

RETROGRADE ORBITS

Retrograde orbits are another hidden problem of current celestial mechanics. The current model hides the problem under piles of math, but as with resonances and torques, the gravity-only model has no way to match the data mechanically. It is known that objects in retrograde orbits lose angular momentum and tend to decay. Triton and Phoebe are the two most famous retrograde orbits in the solar system, and both are thought to be in slow decay. Both are also thought to be captured Moons rather than Moons that formed along with their planets.

Current gravitational theory cannot explain how a torque is applied to a body in retrograde orbit to make it lose energy. But we have seen that the Unified Field explains torque with its E/M component

So let us first look at an orbit like that of Triton from the point of view of current theory. The current model believes that some satellites, like Triton and Phoebe, are captured satellites. Captured satellites must have been captured in the way I showed in my chapter on celestial mechanics—by decelerating into orbit. How was this possibly achieved, given the current list of forces and causes of forces? A large body like Triton enters the field of Neptune and *decelerates*? What, exactly, caused Triton to settle into its current orbit? A balancing of instantaneous velocities cannot explain it, since even if Triton happened to intersect its future orbit at exactly the right distance and at a precise 90° angle, many other factors would also be involved. Neither Triton nor Neptune is an ideal body. They both would have had some spin. Just as an example, it is believed that all bodies apply torques to all other bodies (although it is not explained how in current theory). Therefore Neptune must have a rather complex field at all orbits, not just a simple centripetal acceleration. Scientists use this complex field to explain the motions of Neptune's other Moons. If you add this complexity to the real field of Neptune, you see that the odds of Triton arriving with all the perfect counter-speeds and counter-torques, at just the right angle and distance are precisely zero. The field of Neptune must have some ability to resist small deviations and to correct them. Otherwise no body could ever be captured in the first place.

It is true that the orbit of Triton is decaying, so that the orbit is not in fact completely stable. But this is not the question. No field is infinitely forgivable, but orbits show a degree of float that is not in line with current theory. There appear to be constraints on decay and escape far beyond what would be logically expected. **A decaying orbit like Triton's would be expected to fail exponentially.** As Triton lost energy it would fall into a lower orbit. At this lower orbit the acceleration toward Neptune is even faster. To be in a stable orbit at a smaller radius, Triton would have needed to gain energy, or speed up. It has lost energy and gone lower, therefore we would expect a multiplied affect. Instead we see a long slow decay. Once again, empirical evidence directly contradicts the given theory of gravity and orbit.

According to the postulates of current theory, a decaying orbit would be expected to fail exponentially, and therefore very quickly. A decaying orbit would not last a thousand years, much less millions or billions of years. But that is not what we see.

Now let's return to my theory. Let us say that the torque from Neptune works preferentially on the spin of Triton and not the velocity. In this case it would never appreciably affect the orbital momentum of Triton since Triton is so large. It might only affect the angular momentum, which decreases the energy of Triton's E/M field relative to Neptune's E/M field. In this way Triton loses energy but does not lose speed or radius. If this is the case, then we only have to look at the spin of Triton. Once the spin of Triton is stopped by Neptune, it must begin to reverse, since the torque from Neptune is constant. Eventually Triton will gain enough energy to create its own torque against the field of Neptune. At some point this torque will be sufficient to create a slight addition to orbital velocity, at which time Triton will bump itself into a higher orbit. The affect will become additive and eventually Triton will escape.

You may ask how a more energetic Triton turns that energy into orbital velocity. It does so with that resisting E/M field torque. That torque will have a component that is parallel to the orbit of Triton, and this must increase the orbital velocity. Even if we give the torque preferentially to the spin, there must be some point at which this preferential treatment breaks down. That is, once Triton gains some given amount of angular momentum, the torque can no longer be given to spin, preference or no. At that point the tangential component of the torque will begin affecting the velocity. This would explain why major satellites do not impact their primaries. It would also explain why Triton's orbit decays so slowly.

It also explains why so many objects are in prograde orbits. The laws of chance, given current theory, would provide us with many more retrograde orbits than we actually see. We know from Triton and Phoebe and Pasiphae that satellites can be captured, and we must assume that they can be captured in prograde or retrograde. Even if we imagine that inner satellites formed with the planets, we should still see about equal numbers of outer or captured planets that are prograde and retrograde. But we don't. We see only a couple of retrograde orbits. Why?

It can't be explained with current theory, but it is easy to explain with my theory. Over the age of the solar system, most retrograde orbits have been turned into prograde orbits. The orbits of Triton and Phoebe are just very young orbits that haven't had time to turn.

Another thing my theory explains is the Moon's small increase in orbit. The Moon is currently moving away from the Earth at 3.7cm/yr. Now, we know that the Moon is a very old satellite. According to my theory, young or captured satellites would have slowly decaying orbits inward, as the torque from the primary slowed their retrograde momentum or velocity. Eventually they would become prograde and the torque would begin to force the orbit to decay outward. Our Moon has already gone through it period of decay inwards and its relative stability. Now it is in its latter stage, which is a period of slow outward decay. This would mean that no orbit is ever completely stable. All orbits are in some slow transition, either gaining momentum or losing it.

The TROUBLE with TIDES

Tidal theory is one of the biggest messes in contemporary physics. I will start with tides on the Earth, since they have gotten the most attention and the most theory. We know the ocean tides are caused by the Moon, since they follow lunar cycles. But are they caused by the Moon's gravity? Let's look at some numbers. Let's compare the Sun's field to the Moon's field, at the Earth.

a_S = force on the Earth by the Sun
a_M = force on the Earth by the Moon
$a_S = GM_S/r^2 = .006 \text{ m/s}^2$
$a_M = GM_M/r^2 = .000033 \text{ m/s}^2$

You can see that the Sun has a much stronger gravitational effect on the Earth, if we look strictly at field strength. We could have guessed this without the math, since if the Moon had a stronger gravitational effect, we would be orbiting it, not the Sun. If tides are caused by gravity, then it seems like we should be experiencing Sun tides that utterly swamped our Moon tides. By the math above, Sun tides would be about 180 times as great as Moon tides, making the Moon tides invisible. They would follow the movements of the Sun overhead.

Why aren't we experiencing Sun tides that are stronger than Moon tides? According to an article at Wikipedia, which is following the Standard Model and which is reprinted all over the web,

Gravitational forces follow the inverse square law (force is inversely proportional to the square of the distance), but tidal forces are inversely proportional to the cube of the distance. The Sun's gravitational pull on Earth is 179 times bigger than the Moon's, but because of its much greater distance, the Sun's tidal effect is smaller than the Moon's (about 46% as strong).

These are the basic equations, according to the standard model:

$$F_S = GmM_S / r^3$$
$$= 2.4 \times 10^{11} \text{ N}$$
$$F_M = GmM_M / r^3$$
$$= 5.1 \times 10^{11} \text{ N}$$

These equations, as I have simplified them here, don't give the right numbers, but we do get 46%. How was this "inverse cube law" derived? According to a University of Washington website[1],

Tidal forces result from imperfect cancellation of centrifugal and gravitational forces a distance L away from the center of gravity of the system and have the form $F_t = GmML/R^3$

Other websites agree. Here is one that is especially funny, considering everything:

So the gravitational attraction of the Sun is 178 times greater than that of the gravitational attraction of the Moon. But how can this be? We all know the Moon is more effective in producing tides than the Sun. There is a simple explanation for this, and it is not that we have been lied to! It is only the proportion of the gravitational force not balanced by centripetal acceleration in the Earth's orbital motion that produces the tides.[2]

Two major problems here. One, the gravitational force causes the centripetal acceleration. There can be no lack of balance. As for the gravitational and centrifugal forces, although they are caused separately, they cannot cancel, since they both tend to create tides. In fact, most physics books and websites use a summation of centrifugal effects and gravitational effects to create tides on the Moon, as I will show below, since both are tidally positive. That is to say, gravity would create tides even without circular motion, and circular motion would create tides even without gravity. So the two are additive. There is no possible cancellation, in the way that is assumed above. Besides, the Earth is not feeling a centrifugal effect from the Moon, since the Earth is not orbiting the Moon. Even if it were orbiting a barycenter, it still would not be in circular motion about the Moon. Therefore the tides on the Earth could not be an imperfect cancellation of centrifugal forces and gravitational forces, even if these forces were in opposition. There are no centrifugal forces on the Earth directly caused by the Moon, since there is no angular velocity around the Moon.

Secondly, the math above is dishonest. If we look at the Sun/Earth system, then the center of gravity of the two bodies is so close to the center of the Sun that it makes no difference. The Earth has almost no effect on the Sun. Therefore, the distance L is just the radius R, and the equation is the same as

$$F_t = GmMR/R^3$$

That is not an inverse cube law, it is an inverse square law in poor disguise.

A much better explanation of the inverse cube law is supplied by Wikipedia:

Linearizing Newton's law of gravitation around the center of the reference body yields an approximate inverse cube law. Along the axis through the centers of the two bodies, this takes the form $F_t = 2GmMr/R^3$

"Linearizing" means differentiating the equation with respect to R, so that this new equation represents a change in the field, rather than the strength of the field. Despite being weaker, the field of the Moon changes more quickly. This causes a greater difference from center to far or near edge. Another way to express this without differentiation is:

$$a = GM[1/R^2 - 1/(R-r)^2]$$

Where R is the distance between objects, and r is the radius of the gravitating object. They tell us this equation is approximately equal to $a = GM2r/R^3$, giving us an inverse cube law.

It is clear that the differentiating proves that there would be an inverse cube *effect* in the tide-producing differentials, supposing that the postulates of this theory are true. I don't know that I would call it an inverse cube "law", since it does not apply to the field itself. It applies to the differential field. It comes from the fact that tides in a static gravitational field are determined by the rate of change of the field, not by the strength of the field. What I mean by static is that this calculation does not take into account the circular motion of the object in the field. Even objects in straight freefall would be subject to this tidal inverse cube law, as Wikipedia and current theory admits. But the Earth is not in simple freefall around the Sun. It is in orbit. We must therefore add a centrifugal effect to the static effect of the field. Once again it appears that this must take the Sun's effect beyond the Moon's effect on the Earth.

To find out, let us actually calculate a force. We know that the centrifugal force varies in a different way from the centripetal force. The centripetal force gets weaker as you go out, since it must be assigned to the gravitational field. But the centrifugal force, in this case, increases at greater radii. This is because the far side of the Earth in its orbit must have a greater orbital velocity than the near side. To calculate this force we must first find the acceleration of different parts of the Earth using the equation $a = v^2/R$.

$R = 1.4959787 \times 10^{11}$ m
$v = 2\pi R/t$
$t = 365.257d = 31558205s$
$v = 29784.68322$ m/s
$R + r = 1.4960424 \times 10^{11}$ m
$R - r = 1.4959149 \times 10^{11}$ m
v_o = outer velocity = 29785.95147 m/s
v_i = inner velocity = 29783.41297 m/s
$a = .00593008$ m/s^2
$a_o = .005930332$ m/s^2

$a_i = .005929827 \text{ m/s}^2$
$\Delta a = 2.53 \times 10^{-7} \text{ m/s}^2$
So let's show the basic equation for the math above:
$\Delta a = [v^2/R] - [v_i^2/(R - r)]$
$= [4\pi^2 R/t^2] - 4\pi^2(R\text{-}r)/t^2$
$= \omega^2 R - \omega^2(R - r)$ where ω is the angular velocity ($= 2\pi/t$)
$\Delta a = \omega^2 r$

All that work to get the same number we found at first, way above [.006]. I wanted to show you that the circular motion equation generated the same number as the gravitational equation. This is no accident, of course. Measured from the center of the Earth, the two numbers would be expected to be the same, since it is the acceleration due to gravity that keeps the Earth in orbit, according to gravitational theory. The math above just mirrors the math of the differentiated equations. It is the same in form, but not in output, since the centrifugal field varies differently than the gravitational field, as I have said. Here is the equation for the static gravitational field of the Sun at the Earth:

$\Delta a = GM[1/R^2 - 1/(R\text{-}r)^2] = 5.08 \times 10^{-7} \text{ m/s}^2$

Doing the same math for the gravitational field of the Moon at the Earth, we find

$\Delta a = 1.14 \times 10^{-6} \text{ m/s}^2$

These two equations yielded the number 46%, remember. But now we have some more tidal effect to add from the Sun. The total tidal effect from the Sun is now

$\Delta a = 2.53 + 5.08 = 7.61 \times 10^{-7} \text{ m/s}^2$

This does not take us over the effect from the Moon, but it takes our number for **Sun tides up to 67% of Moon tides.** If we correct the mistakes in the math, we can no longer match data.

The Standard Model, as expressed in Wikipedia and elsewhere, adds the centrifugal effect using this equation:

$\Delta a = \omega^2 mr$

ω is the angular velocity, so, according to Kepler's law, $\omega^2 = GM/R^3$. This makes the equation equivalent to the math I used. This term $\omega^2 mr$ gives us half of the value of the first term, $2GMmr/R^3$. In other words, the tidal effect caused by circular motion is half the tidal effect caused by the static gravitational field. On this much we agree, as you can see from my numbers for the Sun [2.53, 5.08]. But the Standard Model goes on to apply the full equation to the tidal effect on the Earth from the Moon:

$\Delta a = \omega^2 mr + 2GMmr/R^3$

This equation is equivalent to my math above:

$\Delta a = \omega^2 r + GM[1/R^2 - 1/(R-r)^2]$

But neither equation is applicable, **since the Earth is not orbiting the Moon.** The first term on the right side cannot be applied, because if you re-expand it, you find that it contains the variable R. Like this, remember:

$a = \omega^2 R - \omega^2(R - r)$

This R applies to the Earth-Moon radius. But if the Earth is actually orbiting the barycenter, then this radius R does not apply in the first term of the equation. We must use the number 4,671 there, not 384,400. To get the correct angular velocity, we must use the correct radius. The current equations use 384,400 for R, but the value should be 4,671. That throws off all the numbers, and prevents them from getting the 46% they desire. If we correct the math, we find 67%, as I showed above. But that doesn't match data.

I have had readers answer: "But we are subtracting the R out of the equation, so it doesn't matter what it is." It matters because the value of v and ω are chosen to match that radius. The Earth wouldn't have the same angular momentum around a barycenter that it would have around the Moon, would it? The current equations use a value for ω that implies the Earth is orbiting the Moon; when, at best, the Earth is only orbiting a barycenter.

This means that all the standard model math fails. The mainstream has been publishing false equations. I assume they know they are doing this, since the holes in the equations are so big. Using the wrong radius is a huge error, one that is difficult to explain away as an oversight. It would have to be an oversight of many decades, involving thousands of specialists. I believe it is a purposeful fudge.

[Note added August 2007: Confronted with parts of this paper in late 2005, Wikipedia deleted all its tidal theory math, its tidal theory page, and ordered a rewrite with lots of new illustrations. It appears they are perfecting their propaganda rather than admitting that their math and theory doesn't work. This change affected many other websites as well, since Wiki is linked to a large percentage of online encyclopedic entries. Large parts of tidal theory have gone into hiding since the publication of this paper. One place that is still hanging the dirty laundry out in the open is the department of Oceanography at Texas A&M (ocean.tamu.edu), as I was informed by a reader. All the tidal math there falls to this critique, since it is equivalent to the math that was up at Wiki. It is a pretty variant, but it includes the inverse cube law above, and uses the wrong radius. It is very clever at hiding all the problems, since it hides some variables and refuses to define or assign others. It also hides the barycenter problem, revealed just below.]

Another major problem with tidal theory concerns its use and misuse of the barycenter. The barycenter is the center of gravity of the Earth/Moon system, which both bodies are said to orbit. Feynman was one of the most famous to suggest that the Earth has a non-negligible tide created by orbiting this barycenter. Is this true? Let's do the full math.

$R = 4671$ km
$v = 2\pi R/t$
$t = 27.32d = 2360448s$
$v = 12.43$ m/s
$R + r = 11042$ km
$R - r = -1707$ m
v_o = outer velocity = 29.39 m/s
v_i = inner velocity = -4.54 m/s
$a = 3.31 \times 10^{-5}$ m/s^2
$a_o = 7.82 \times 10^{-5}$ m/s^2
$a_i = -1.2 \times 10^{-5}$ m/s^2
$\Delta a_o = 4.51 \times 10^{-5}$ m/s^2
$\Delta a_i = 4.51 \times 10^{-5}$ m/s^2

We certainly do find a significant effect from the Earth orbiting its own barycenter. In fact, it swamps all other effects. It is 40 times as great as the gravitational effect from the Moon and almost 60 times the total effects from the Sun. However, Feynman was wrong in one very important way. The effect doesn't just raise a tide on the far side of the Earth from the Moon; it raises an equal tide toward the Moon. Feynman obviously didn't know what to do with that negative radius. But as you can see from my diagram, it produces a positive tide. You must follow the steps of the math I did previously, and if you do it exactly, you find that you must subtract a_i from a, to achieve the proper differential. As vectors, they are pointing in opposite directions, so you subtract a negative, which is the same as addition: $\Delta a_i = a - a_i$

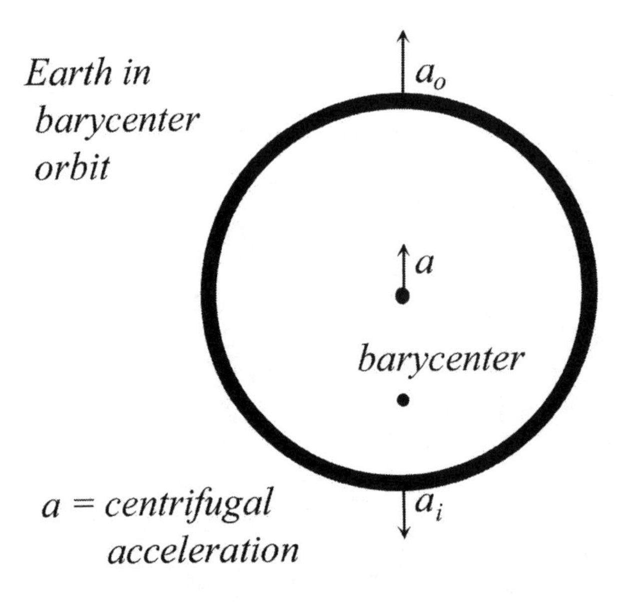

The barycenter falsifies the entire standard analysis, since it would swamp all effects from the Sun and Moon. You cannot include effects from the barycenter, since they cannot be made to fit the given data. And you cannot fail to include effects from the barycenter, since current gravity theory demands a barycenter. This is called a failed theory.

Some have tried to squirm out of this by telling me that since the barycenter numbers are equal forward and back, the tide is a constant and the other tides can just be stacked on top of it. But this is illogical. The numbers show the barycenter tides equal front and back: they do NOT show an equal tide all the way around. The barycenter is a discrete point, and there is a vector pointing from the center of the Earth to the barycenter. This gives the barycenter tide position and direction on the surface of the Earth, just like any other tide. In other words, it would create heaps. That is what my illustration shows. The math and the illustration do not show an equal heaping in all directions. All points on the oceans would not swell equally, and at the same time. Therefore, if we had a barycenter tide, it would be quite obvious. It would create huge swells and it would swamp all other tides. We wouldn't need a list of 388 tidal harmonics, we would need only 1.

Part 2: Tidal Effects on the Moon

Now let us look at tides on the Moon. I will start over with my analysis, pretending once again that the reader knows nothing about tides; but in this section I will hit some topics that we missed in the first section. The Standard Model, as glossed in textbooks at all levels, explains tides by showing that real bodies do not behave like point particles. Because they have real extension, different parts of the body must be feeling different forces. If we take the Moon as an example, we can compare three points on or in the Moon. We take the point nearest the Earth, the point at the center of the Moon, and the point farthest away. The point at the center feels a force from the Earth that is just sufficient to make it orbit. That is why, in fact, it is orbiting. It feels no tides of any kind. The point nearest the Earth requires less force to make it orbit than the point at the center, but it actually feels more force. The point farthest from the Earth requires more force to make it orbit, but it is feeling less force than the point at the center. The point nearest therefore feels a resultant force toward the Earth and the point furthest feels a resultant force away from the Earth. This causes a tide that maximizes at the near and far points.

So far so good. The Standard Model applied to the Moon follows what we have already found regarding the Earth. But before we analyze it again, let's look at something interesting. Notice how theorists who claim to believe in General Relativity always revert to Newton when it comes time to explain forces in gravitational fields. In the chapters on General Relativity we are told that an orbiting body is feeling no forces. It is simply following curved space, the "line" of least resistance. We are shown the ball-bearing on the piece of rubber, and the tiny marble orbiting it with no centripetal force. All quite ingenious, except that it does not explain the genesis of the forces at a distance used in tidal theory. How can an orbiter that is feeling no forces achieve tides? Even more to the point, how can an orbiter that is traveling in the curved space of its primary re-curve that space in order to transmit a tidal force to the primary? Is the gravitational field between the Moon and Earth curving convex or concave, relative to the Moon? I would think it must be one or the other. It cannot be curving both ways at once.

If anyone answers "gravitons," then I think we can throw out the curved space idea as superfluous. If we have gravitons mediating the force, then the Moon is feeling a force. In which case we don't need curvature to explain anything.

But the current theory isn't even that advanced, regarding tides. Graviton or no graviton, the theory reverts to Newton for the explanation. To cover all its bases, the theory gives the situation a sort of double cause. The first cause is given to the gravitational field. Nearer parts of the body will accelerate toward the Earth faster than farther parts, regardless of their weight or mass. Remember that acceleration in a gravitational field has nothing to do with mass. All objects fall at the same rate. Acceleration is dependent only on radius. So the analysis should always be talking about accelerations, not forces. This part of the theory is at least logical, given Newton's equations. It is true that the static gravitational field would create tides as claimed, near and far. But it would create these tides even if there were no circular motion and no orbit. An object in freefall would experience this sort of tide, as the Standard Model admits.

Current theory gives a second mechanism, and this mechanism requires an orbital velocity. Nearer parts of the Moon orbit in a slightly smaller circumference than farther parts. They travel this circumference in the same time as the rest of the Moon. Therefore they have a slower orbital velocity. With more acceleration and less orbital velocity, the near tide is increased. Likewise, farther parts of the Moon have less acceleration and more orbital velocity, once again increasing the proposed tide. This analysis is once again (mostly) true, but this second cause has nothing to do with gravity. It is an outcome of all circular motion, whether you have a gravitational field or not. Whirl any dimensionally consistent object and the circular motion will create tides in the object just like these, if you apply the forces in the same way.

So, we can add up the effects on the Moon just like on the Earth. We can use the equation

$$\Delta a_E = \omega^2 r + GM[1/R^2 - 1/(R\text{-}r)^2]$$
$$= .000012 + .000024 = 3.6 \text{ x } 10^{-5} \text{ m/s}^2$$
$$\Delta a_S = \omega^2 r + GM[1/R^2 - 1/(R\text{-}r)^2]$$
$$= 6.9 \text{ x } 10^{-8} + 1.4 \text{ x } 10^{-7} = 2.1 \text{ x } 10^{-7} \text{ m/s}^2$$

According to current theory, the tidal effects from Earth to Moon are 32 times the tidal effects from Moon to Earth. And according to these same equations, the solar tide on the Moon should be 171 times smaller than the terrestrial effect. More importantly, the visible tide on the Moon should be symmetrical front and back, as caused by the Earth. Is this what we find? Not at all. The Moon rotates relative to the Sun, so we would not expect to find a solar effect on the Moon, beyond a tiny constant shift in the crust opposite the direction of this rotation. The rotation of the Moon on its axis relative to the Sun does not cause a further tide from the Sun, or add to the tidal effect, but it acts to shift the tides we have already calculated, just as the rotation of the Earth shifts the ocean tides, causing them to travel. I am not aware of any experiments on the Moon to measure lateral shift of the crust in the direction opposite rotation, to verify the relative strength of the solar tide, although this would be a very useful experiment. However, concerning the terrestrial tide, we have ample visual data. This data is not a confirmation, to say the least. A schematic of the Moon from any textbook will show you that the Moon exhibits a much greater tide at the back. In fact, it has a negative tide at the front, the crust being almost obliterated in places.

How does current theory explain this? It can't explain it using gravity or circular motion. This is how it is explained in the Encyclopedia of the Solar System:[4] "The conventional explanation for the center of figure/center of mass offset is that the farside highland low-density crust is thicker. It is massive enough and sufficiently irregular in thickness to account for the effect." More gobbledygook, in other words. If the farside crust is low-density, this would only add to the problem. To create a greater tide we need more mass over there, not less. And farside variations do nothing to explain the nearside obliteration.

Before I close, I have one more thing to say about the orbit of the Moon. In all these analyses, both mine and those of the Standard Model, it has been assumed that outer parts of the Moon can travel faster than the inner parts. The diagram requires it and so we have just taken it as a given. We do not even ask how it is physically possible for different parts to have different tangential velocities and different orbital velocities. The gravitational field cannot be creating them, since it cannot exert a force tangentially. The field creates only radial forces. We need either a mechanical cause of the variance, or we need to show that all orbiters exhibit shearing along the direction of orbit. Orbiters in tidal lock should exhibit strong symptoms of shearing, since the forward part of the object is always in lower orbit and the back part is always in higher orbit. The back part of the Moon should shear in the reverse direction of orbit and the front part should shear in the forward direction. But the data is negative, and we are given no cause for the negative data. The only mechanical cause would be some sort of absolute rigidity of the Moon radially. But this is not true empirically. With current theory, the lack of data is a complete mystery.

General Relativity can explain it, since according to that theory, the Moon is feeling no forces. A Moon feeling no forces would not be showing any signs of shearing. But you can hardly use GR to explain the tides we don't see and use Newton to explain the tides we do see. The Moon is either feeling forces or it isn't.

The problem was a big one for Newton, even in his own time, since he is the one who postulated that the tangential part of the velocity in orbit was caused by the orbiter's "innate motion". That is to say, the tangential vector is one the object has prior to or independent of the gravitational field. But of course the object could not have a variable innate motion. It cannot speed up outer parts and slow down inner parts just to suit diagrams.

It is now not just Newton's problem. Current theory has inherited it and failed to explain it, or even try to explain it.

[1] www.npl.washington.edu/AV/altvw63.html. This site is managed by CENPA, the Center for Experimental Nuclear Physics and Astrophysics.

[2] www.sanho.co.za/tides/tide_theory.PDF

[3] www.co-ops.nos.noaa.gov/restles1.html#Intro

[4] Compiled by NASA and JPL, 1999. p. 252.

The MAGNETOPAUSE CALCULATED from the UNIFIED FIELD

It has been found that the Solar Wind works differently with positive ions and negative ions.* Protons are accelerated by the Wind in an even manner, passing the Earth in numbers and at velocities that can be predicted from various models by the temperature of the corona. But electrons behave in an unpredictable manner, not being accelerated at the proper velocity. They are moving too slowly. They have also been found to be diverted by magnetic field lines, while the protons were not.

This phenomenon, though long known, has never been explained. The standard model cannot explain it because the dipole field of electromagnetism is supposed to be balanced. That is to say, the proton is not given more charge than the electron, or vice versa. The electron attracts the proton just as much as the proton attracts the electron. Given a field of potential like this, there is no way to explain the different behavior of negative charge and positive charge. According to the standard model, the solar system is nearly neutral as a whole, so ions are accelerated due to very limited field or no fields. In other words, they are not being accelerated by some long-range potential between the Sun and outerspace, they are being accelerated by short-range potential differences in the outer layers of the Sun, or are being ejected directly from the interior as thermonuclear by-products. Other explanations are also advanced. But none can explain the data. If the ions are accelerated by charge, then the electrons should be going in opposite directions to the protons. If they are accelerated by mass, the electrons should be accelerated more, not less. The proton has more inertia, so it should resist acceleration better. Likewise if ions are ejected from the solar interior: electrons should be ejected at greater velocities, since they are smaller. Or, if they are both ejected near c, and are nearly equivalent due to the limit at c, then they should be nearly equal. In no case should the electron be accelerated less, or be more easily diverted.

But the plasma or electrical Sun model can also not explain it. The only prominent competing theory of solar energy to attract any attention in the past half-decade has been the electrical model, which came (in twisted channels) from Velikovsky and plasma research. According to this theory, the Sun is a giant anode being fed energy from cathodes in the rest of the galaxy. But if this is so, then only the protons should be accelerated out from the Sun. The electrons should either never be ejected, or they should loop back immediately. This theory is also contradictory in the way it treats the Earth and the Solar Wind. According to Ralph Juergens (following Tesla), the Earth acts like a well of negative charge. As such, it should repel negative ions in the Solar Wind. Instead, we find that the Earth, via its magnetosphere, excludes both negative and positive ions. The E/M field, which is supposed to be field of potential in both the standard and plasma models, is not acting like a field of potential.

Both models use plasma to explain Solar Wind exclusion, but neither model is consistent. Let's look at how Wikipedia uses plasma to explain Solar Wind exclusion. On the page entitled "Magnetosphere", we are told of the Solar Wind that:

Its composition resembles that of the Sun—about 95% of the ions are protons, about 4% helium nuclei, with 1% of heavier matter... and enough electrons to keep charge neutrality.

See a problem there? You cannot maintain charge neutrality with 99% positive charge. That leaves less than 1% negative charge, and <1% cannot balance >99%. The electron and proton have equivalent charges, by the first postulate of modern theory. Then we are told,

Physical reasons make it difficult for solar wind plasma with its embedded [interplanetary magnetic field] to mix with terrestrial plasma whose magnetic field has a different source. The two plasmas end up separated by a boundary, the magnetopause, and the Earth's plasma is confined to a cavity inside the flowing solar wind, the magnetosphere.

"Physical reasons." I will have to remember that next time someone asks me a question about mechanics. "Physical reasons," I will say. I am not questioning that plasmas may create these boundaries, I am pointing out that we require a mechanical explanation for it. An existential explanation will not do. "Because plasmas work like that" is not a mechanical explanation.

Since neither the standard model nor the plasma model has given us a satisfactory explanation for the electromagnetic action of the Solar Wind, I will offer a third model here, one that is far simpler and far more comprehensive.

In a series of other papers, I have shown that the Solar System is neither wholly gravitational, as the standard model would have it, nor mainly electrical, as the plasma model would have it. Like all else in the universe, the Sun and its environs are driven by the unified field. I have shown that Newton's main gravitational equation is really a simple unified field equation, and that it has contained the charge field from the beginning. I will now show how this explains various electrical anomalies in the Solar Wind and in the electrical fields of Solar System bodies.

To begin my explanation, I remind you once again that the charge field in my mechanics is not equivalent to either the electrical field or the magnetic field. The charge field underlies the E/M field, but is not strictly equivalent to it. The E/M field is a field of ions, but the charge field is a field of real charge photons—not virtual or messenger photons, but real particles with mass and radius and spin. These charge photons work strictly mechanically, by bombardment. All material particles (except photons) emit a steady stream of charge photons, to create the charge field. Since the Sun is composed of a stupendous number of material particles, it emits a stupendous charge field. This charge field is emitted spherically or radially and is, in the first instance, always repulsive. The charge field is monopole: its velocity and momentum is always radially out from the center.

It varies in only one way: the charge photons may be upside-up or upside-down. In other words, the photons may be spinning either left or right. As a matter of convenience and symmetry, we may call the left spinners anti-photons. But both photons and anti-photons are emitted in the same direction. They are not dipole. They differ only in spin, not in field potential or linear motion.

"Where do these photons come from?" you may ask. "Shouldn't an emission field with mass cause a conservation of energy problem?" It doesn't, because the field is recycled. All the masses that emit the charge field also absorb or capture the charge field. It is the spin of the mass that allows it to do this. As I have shown with the atomic nucleus, the spin of the particle or collection of particles creates a low pressure at the axis poles. The charge field goes in at the poles and is ejected equatorially. All matter is an engine that exists by recycling this charge field. As with the proton and nucleus, so with all macro-spheres. Spin creates pressure variations in the charge field, which creates the simple engine. The Sun (and Earth) capture the charge field at the poles, and re-emit it everywhere else—but most along the equator.

The electromagnetic field is driven by the charge field, by direct bombardment. At first glance, you would think the electrons and protons would be driven equally, or that the electrons would be driven faster because they are smaller. But on looking closer, you see that the size difference between the electron and proton causes just the opposite effect, in a simple mechanical way. The electrons are driven less by the photon wind, because they can dodge greater parts of it than the proton. The radius of the electron is some 1800 times less than the radius of the proton, so large parts of the photon wind simply miss it.

Therefore, the proton is driven more efficiently. This explains in a direct manner the data from the Solar Wind.

It also explains the deflection of the electron by planetary or Solar magnetic fields. Magnetic fields are caused by the spins on the photons, not by the linear momentum. The proton feels more of the linear field, since it gets hit more often, but it resists the spins of the photons better because it is larger. The protons and electrons are also spinning and are also emitting small charge fields of their own. But, contra the standard model, the proton actually has a greater charge than the electron, simply because it has a greater radius and therefore a greater angular momentum. This greater angular momentum allows it to resist the much smaller angular momentum of the charge field. The electrons, although hit less often, are more likely to be deflected (as a statistical matter on individual ions), because they feel a much greater relative force from the angular momenta of the photons.

In this way, the Sun is both anode and cathode, but only as regards the charge field. Due only to pressure differences, it attracts the charge field at its poles and emits the charge field everywhere else. You can now see this with your own eyes by watching a NASA film called The 3D Sun.† At minute 19:43 you will see the heaviest emission near the Solar equator and the lightest at the poles. If the Sun were spinning faster, this effect would be increased.

On a smaller scale, this also applies to the Earth. The Earth's spin makes it both anode and cathode to the charge field. It recycles the charge field, and the charge field drives the E/M field. This explains the genesis of the Earth's E/M field without postulating dynamos in the Earth. This also explains why the Earth, like all macro-bodies, often seems to be an infinite well of negative charge. Neither the standard model nor the electrical/

plasma model can explain why the Earth should act like an infinite well of negative charge. For example, it absorbs a huge amount of mainly positive cosmic rays each year for billions of years with no drain. My unified field theory explains it by re-defining charge. The Earth recycles both protons and electrons, so both forms of charge are continuously renewed. There is no dipole, so the amount of one charge does not deplete the potential of the other. In fact, there is no potential at all, except the real pressure difference in the charge field, and the apparent electrical difference caused by the size differential between the electron and proton. And the magnetic field is not caused by potential either. It is caused by spin. In the magnetic field, quanta aren't turned by potential differences, they are turned by angular momentum.

This re-definition of the charge field and thereby the electromagnetic field resolves all at once the Velikovsky affair and the role of E/M in celestial mechanics. In fact, it solves all the problems of celestial mechanics, all the way back to Halley and Laplace.

Astrophysicists in the 50's and 60's could not accept Velikovsky's claims for electrical perturbations between solar system bodies, and physicists of all sorts still cannot accept any evidence of large electrical influences from anyone, and this is because neither the outsiders nor the insiders have been able to say how these influences can fit into equations of celestial mechanics—equations that already work almost perfectly in most situations. Shapley in the 40's and 50's could not countenance any pervading E/M field in the solar system for any reason, because he knew the equations of Kepler and Newton and Laplace already worked. There was no room for an addition. Sagan made the same arguments in the 70's and 80's, and current mainstream physicists are holding firm to that line. This is quite understandable, since the plasma physicists still haven't shown them where to fit the E/M field into the current successful equations. Velikovsky was full of ideas, but he never supplied a single line of convincing math.

But my simple equations do just that. I show where the E/M field fits into Newton's gravitational equation. The E/M field is already in the equation, hidden by G, so no external correction is necessary in most situations. Likewise, I have shown that the E/M field is already inside Kepler's ellipse equation and inside Laplace's great inequality equations. It has never been a matter of overthrowing celestial mechanics, or of creating a new unified field equation. I have shown that it was always just a matter of understanding that Newton's equation was already a unified field equation. Because Newton's mass field was already a unified field, Einstein's field equations are already unified field equations, too. We did not need to unify, we needed to segregate, so that we could see how both fields create the current equations mechanically.

If you have not yet read my analysis of Newton's gravitational equation, none of this will be plausible, so I encourage you to do so. But those readers who have will understand that this explains the boundary between the Earth and the Sun, the so-called magnetopause. The boundary between the plasmas is caused mechanically by the charge field, since the Earth's charge field is what creates and defines its plasma to begin with. At the foundational level, the boundary is the boundary between the charge field of the Earth and the charge field of the Sun. Both field are summed radially out, so we simply have a meeting of two spherical fields, one much smaller than the other. Since the charge field is always repulsive, it will exclude all ions, positive and negative, unless the ions are too energetic to be excluded (as with cosmic rays).

This is why Venus can also exclude the Solar Wind, even without a magnetosphere. The ability to exclude has little to do with the magnetic field and more to do with the charge field. Venus has enough mass to create a charge field strong enough to exclude the charge field of the Sun at that distance. With the Earth, the same is true. It is not the magnetosphere that excludes the Solar Wind, it is the charge field. The charge field creates both the magnetosphere and the boundary, and the boundary is only at the edge of the magnetosphere because both the charge field and the magnetosphere pause at the same place, for the same reasons. The magnetopause is at the charge-pause, because the magnetic field is caused by the charge field. The magnetic field would not be expected to extend beyond the charge field, because the charge field creates the magnetic field. And the magnetic field would not be expected to stop short of the charge-pause, because there is no reason for it to stop short. Ions are driven up by the charge field, and will persist as long as the charge field persists.

By this logic, you can see that it is the difference in the charge field densities that creates the differences in plasma fields, and the boundary. The boundary is that distance where the two charge fields have equal power. They have equal exclusion—defined as linear momentum—but one has more curvature than the other. This means that the magnetic or orthogonal components are not equal, creating a boundary and a "sideways" force. So the Solar Wind doesn't stop or blowback, it goes around. Since the Earth's field has more curvature, the magnetic component at the boundary acts to accelerate the Solar Wind once more. This is why the Wind is going faster as it passes close to the Earth than it was before it got there.

Wiki tells us that the magnetopause is 10^{-12} Earth radii in the direction of the Sun. Using my mechanics, we can calculate that distance directly—something the standard model never does and cannot do. The mass of the Sun is 332,990 Earth's and its density is .255 Earth's. We seek a charge density on the surface of the Sun, and we can get that by just looking at the words. We seek a "charge density". That could be written "charge x density", and, as I have shown, charge is just a variant definition of mass. Therefore, we re-write the product as "mass x density." M x D = 84,986. The Sun's charge density is 84,986 times that of the Earth. So we find by simple math that the charge field density is vastly different than the material density. We are finding the charge field density at the surface of the Sun, so we must sum all the mass "behind" that surface. All that mass emits charge photons.

Now we seek the point between Earth and Sun where the two charge densities are equal. Since I have proved elsewhere that the charge field, when emitted by spheres, diminishes as $1/r^4$, we can solve. If the Sun's relative charge density is 84,986 at 1 Sun radius, at 214 Sun radii it will be .00004052. If the Earth's density is 1 at 1 Earth radius, at 12.53 Earth radii it will be .00004057. That is where the charge field strengths match. Since Wiki is measuring from the Earth's surface, we must subtract one Earth radius, giving us 11.53 Earth radii for the charge-pause.

The mainstream cannot supply this simple math because their E/M field has no mechanics underneath it. They are still trying to solve with potentials, which are forces at a distance. The standard model likes to denigrate Newton for having a gravitational field containing force at a distance, but modern E/M theory still is based on force at a distance, even in QED. QED has no foundational mechanics, except a pseudo-mechanics based on the virtual messenger photon. Virtual particles are even more brazen cheats than "physical reasons," since "physical reasons" is just a dodge. Virtual particles are a non-mechanical lie, told straight to your face.

I encourage you to seek any mainstream math for the distance of the magnetopause. There is none because there can be none. The estimate from Wiki quoted above of 10^{-12} Earth radii is based on orbiter data, not on math or mechanics. It has to be rough because the orbiter data is rough. But, in fact, there is almost no variance in the magnetopause average. The magnetopause is dependent on the charge-pause, and the charge-pause is determined by the vital statistics of the bodies. Unless the Earth's or Sun's masses or radii change, the charge fields will not change. The magnetic fields can change, due to the ions present, but the charge field cannot change. For this reason, the magnetopause must always return to the average set by the charge-pause, and this average is solid. It is a number that can be found easily, as I just showed.

We can do the same math on Venus, showing that the smaller charge-pause is caused by the nearness of the Sun, not by the lack of a magnetosphere. The mass of the Sun is 408,589 Venus' and its density is .2706 Venus'. If the Sun's relative charge density is 110,548 at 1 Sun radius, at 153.6 Sun radii it will be .0002. If Venus' density is 1 at 1 Venus radius, at 8.4 Venus radii it will be .0002. So the charge-pause of Venus is 7.4 Venus radii from the surface, in the direction of the Sun. That is assuming the low angular momentum of Venus does not further retard its charge field emission. There is some evidence that the speedy and heavy atmosphere of Venus may help in charge field capture at the poles**, but I don't yet have a mathematical expression for the charge field that includes spin. Notice, however, that the standard model has no number, mathematical or from data, for the ionopause distance of Venus. I have just estimated it with simple math. This gives me a chance to make another prediction. I predict that the ionopause of Venus will be found at near 7.4 Venus radii. If it is less than that, I predict it will be found to be because Venus has very little spin.

*Ogilvie, Journal of Geophysical Research, vol. 76, no. 34.
**The Venus Express probe, launched by the European Space Agency in late 2005, has discovered a huge double atmospheric vortex at the south pole.
†http://www.hulu.com/watch/81732/3d-Sun

HOW to BUILD a NUCLEUS without a STRONG FORCE

Contemporary physics has sold us a quantum interaction called the strong force, which is supposed to be one of the four fundamental forces of nature. This interaction has been forced down our throats despite the known fact that there is no evidence for it. The strong force is just a theory. It is a theoretical force proposed to counteract E/M repulsion in the nucleus. Problem is, quantum physicists have never proved that there *is* an E/M repulsion in the nucleus. They have simply assumed that there is. Because the E/M field is known to be ubiquitous at the macro-level, quantum physicists have assumed from the very beginning that it must be present in the nucleus. If it is present, it must be overcome, to explain the nearness of nucleons to one another.

But I will show that the E/M field is not present within the nucleus. This means that the strong force is one more theoretical and mathematical ghost.

To prove this, I will actually show you how to build a nucleus without the strong force. I will diagram several nuclei, showing that it is quite easy to construct a model of protons and neutrons in which all E/M emission is external.

To achieve this requires only a few simple postulates. The first postulate is that the E/M field is caused by an emission field. Protons must be emitting something in order to create the repulsion. The standard model already accepts this postulate, in a mutated and illogical form. The standard model proposes that charged particles emit virtual or messenger photons which are capable of telling other quanta to move away or move closer. Since this proposition is non-mechanical, non-physical, and magical, and since my papers are concerned with physics and mechanics, I must bring this proposition back in line with logic and physics. To do that, I only have to get rid of the "messenger" or "virtual" part of the theory. I propose that protons emit real photons, and that these real photons cause repulsion by simple bombardment. Since we are dealing with protons and neutrons here, but not electrons, I do not have to explain charge attraction. When speaking of nuclear forces and interactions, charge attraction is not a concern. However, I have explained charge attraction in many other places.

The second postulate is that quanta are spinning. Again, the standard model already accepts this postulate in a mutated and illogical form. Quanta are given various quantum numbers called spin or angular momentum and so on, but then the standard model denies that quanta are spinning physically. We are told that they are point particles, and that the numbers are mathematical in nature, not physical in nature. As a matter of physics, this must be a strange assertion, but it is an axiom of contemporary physics. Students of physics are warned, in very serious tones, not to try to diagram or imagine anything at the quantum level. We are told that the quantum arena is inherently strange and mysterious, not amenable to logic or reason. As a matter of job protection, that is an understandable warning; as a matter of physics, it is just silly.

I return to logic, and logic states that if quanta exist, they have extension. If they have extension, they may have real angular momentum or spin. This angular momentum can then be analyzed just like angular momentum at the macro-level. This applies to all quanta, even the smallest. Photons have real spin just like planets or stars or galaxies. But here, we only need postulate that the nucleons are spinning. We know that nucleons have appreciable size, especially compared to photons, so that postulate is not difficult to make. Nucleons are huge compared to photons or electrons, so why not diagram them with spin?

In fact, I have already shown that baryons (protons and neutrons) have four stacked spins. These stacked spins are fully capable of explaining all the characteristics now given to quarks, without a quark model. It is these spins which will allow me to build the nucleus without the strong force.

To begin, we will look only at the outer or z-spin of the baryon. The proton and neutron are both spinning, and since they are approximately the same size, their z-spins will have approximately the same angular momentum. What makes the two particles so different is that the proton is emitting a charge field and the neutron is not. The neutron is swallowing its charge field, since the photons cannot navigate the maze of spins. The four spins of the neutron bring the photons back to the center, while the four spins of the proton allow the photons to escape. I have diagrammed this in previous papers. What this means for our analysis here is that the proton must be treated as an extended particle, while the neutron is treated as a discrete particle. In other words, in this first part of the analysis, the neutron is treated mainly as a z-spin, while the proton is treated as a z-spin plus the shell of emitted photons.

To visualize this easily, think of a lawn sprinkler, one that spins like a pinwheel. The neutron is like the lawn sprinkler spinning, but without the water. The proton is like the lawn sprinkler plus the water being emitted. In this way, the proton acts like a much larger particle, and that is how we will diagram it. You see, to make sense of the nucleus, I must diagram both the particles and the charge field. To do this, I will have to give the charge field both a presence and a direction at all points.

As the next step, we will have to represent the nucleons in some simplified way. Since the nucleons are spinning, we may simplify the sphere into a circle. To justify this, I will ask you to first imagine a sphere emitting a field in all directions. Then, let this sphere spin about a N-S axis, like the Earth. Due to centrifugal forces, nearly all the emission will now be moving in a direction out from the equator. Almost none will be emitted N-S. To put it another way, the emission field of this spinning sphere will have developed large holes at the north and south poles. If the emission field is a charge field, then the charge field will have large minima in the north and south directions. In fact, the spinning sphere will act very much like a spinning disk, with most of the charge being emitted equatorially. The faster the sphere is spinning, the more it will act like a disk. For this reason, in a simplified diagram we can treat the proton as a disk. As a matter of its charge field influence on surrounding bodies, the proton acts much like a circle or disk, which helps when we need to diagram it on a piece of paper.

The same analysis can be applied to the neutron, since it is also spinning. The z-spin causes most of the angular momentum of the neutron to be expressed equatorially, so we can also treat the neutron as a circle.

With these postulates in hand, we are now ready to look at the simplest nucleus with more than one nucleon: helium. We know that helium has an atomic weight of about 4, with two protons and two neutrons, but the standard model has never told us clearly and distinctly why that is so. Why doesn't helium just have two protons? Or, why not 1 neutron or 3 or 4 or any other number? Why is the atomic number 4 stable? The standard model avoids questions like this with much misdirection and disinformation, but I can answer it with very simple mechanics and clear diagrams.

Since I have already shown that the spinning protons must have large charge holes north and south, we simply let the protons meet hole to hole. When helium is created by pressure (as in a star or Big Bang), these holes naturally align. Once the pressure is turned off—when the matter escapes the star—the alignment either persists, because it is in a stable form, or it decays, because it is not.

Two protons hole to hole align naturally under pressure, since this is the lowest energy state, but once the pressure is turned off, the two protons are again free to turn. If the two proton disks start to turn, the two charge fields hit each other and we have repulsion. The two protons together are not stable, even when originally aligned hole to hole, since there is nothing to prevent them from drifting and turning. But if neutrons are present in the star, and if they happen to be present in the right places, they can provide this stability. This is quite easy to diagram:

Helium nucleus

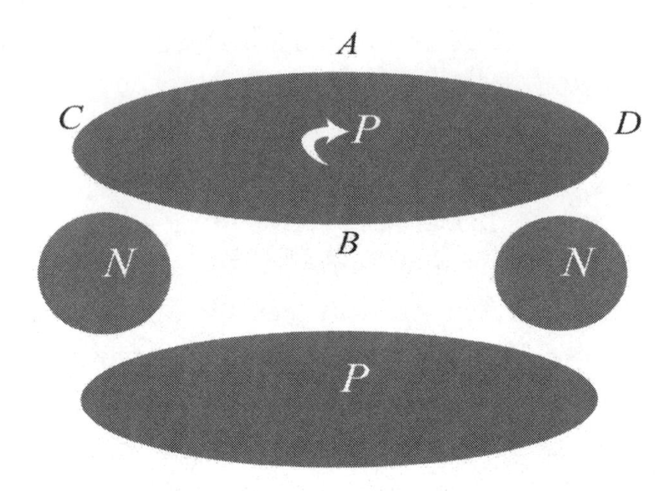

As you see, the neutrons act like little pillars or posts, keeping the proton disks from turning and repulsing one another.

You will say, "Don't we need at least four posts to keep the protons from turning toward each other? Can't the proton disks still fall together at points A and B?"

No, we don't, and they can't. You would be correct, if the disks were not spinning. But because they are spinning, stability can be created by only two posts. For the disks to fall together at point A, say, would require a point on the edge of the disk to be high at point C, low at point A, and high again at point D. To do this, the disk would not "fall", it would warp. We have simplified our sphere into a disk, not into a floppy hat.

My assumptions here are very normal, and you can test them in real experiments with real disks. Spinning disks can be kept apart, in permanent stability, by two posts, as I have diagrammed it.

Since that is true, I have already proved my point with regard to helium. As you can see from the same diagram, no charge is being emitted into the nucleus. All charge is being emitted to the outside. We may assume that some tiny fraction of charge force is being felt by the neutrons, although they are orthogonal to the main line of charge. They could not provide stability if they did not resist turning, and they could not resist turning except by mechanical means. But we will assume that the angular momentum of the neutrons is more than enough to offset this small amount of charge encountered. Using commonsense mechanical postulates, we see that the bulk of the charge is emitted outside the nucleus. Therefore we have no need for the strong force in the helium nucleus.

We can apply the same analysis to lithium. We have three protons and four neutrons. We stack our three disks, and need four posts to separate them.

But now we arrive at the beryllium nucleus. In this case we have four protons and five neutrons. Why that number? Why is the number 9 stable when the numbers 8 and 10 are not? If we use the same diagram as we used for helium and lithium, we would expect to need 6 neutrons to separate 4 protons, which would give us 10. Obviously, the nucleus has already discovered a more efficient method than our dual posts. Beryllium 10, with 6 neutrons, is actually very stable, with a half life of over a million years, so nature does use the six post model here. But the five post model is also effective, so given the chance, nature will prefer it. Beryllium can stack with only five posts due to the fact that the lithium model is already so stable. If we place the neutrons in lithium like this, then we have such a solid spinning structure that the top level can be balanced by only one neutron, placed in the middle. The disk below cannot turn, so the central neutron must resist only the upper disk. Remember that the neutron is not a narrow pillar. It has a z-spin radius equal to that of the proton, so it is quite capable of providing stability in this way. If we let it spin in the same plane as the protons, this is even more obvious.

You will say, "Well, if we can balance disks so easily, why did we not let one neutron balance the third proton in lithium? Weren't the first two disks almost as stable?" Yes, they were, and we can. Lithium 6 is a stable isotope, existing abundantly in the universe. The reason it isn't as common as lithium 7 is probably due to the fact that it is burned more easily in stars. It is slightly easier to break that one post than the two posts of lithium 7, so stars will burn lithium 6 preferentially.

Beryllium nucleus

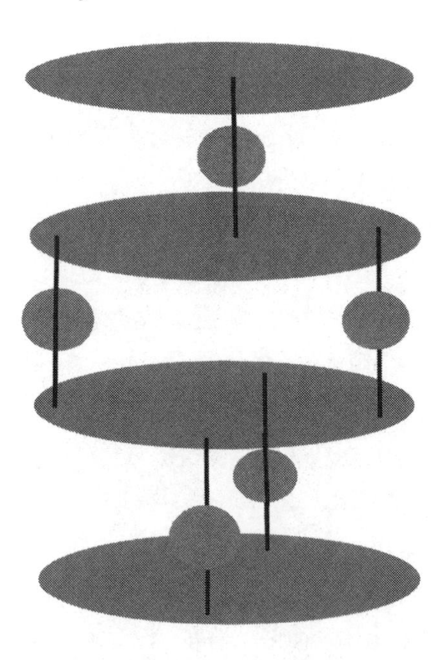

The same analysis applies to helium 3. Helium 3 is stable, but easier to burn than helium 4.

Our stacking method also explains boron, since we use the one post top and bottom, giving us five protons and six neutrons. As expected, boron 10 is also stable, but it likes to capture an extra neutron to achieve even more symmetry and stability.

With all the smaller elements, this disk stacking is both simple and intuitive. And, as you can see, it continues to keep the E/M field out of the nucleus, even as we go down the periodic table.

Now that we have a new model of the nucleus, we find that many things are different than before. Not only have we bypassed the need for a strong force, we have completely overturned the old model of the nucleus as a roundish collection of nucleons, crammed together willy-nilly. The nucleus is not a formless conglomeration, like a bag of marbles, it is a well-defined stack, with many rules of stacking. Beyond that, the nucleus can no longer be considered a simple ion, seeking electrons. It must now be considered a very complex engine. Why is it an engine, you may ask? Well, notice that the stack of protons still has a hole top and bottom. It has charge minima at both ends of the stack. These holes act as intake valves, through which the nucleus can capture other quanta. We may imagine that it can capture anything from photons up to electrons, and possibly even larger quanta. What does it do with these captured particles? It "burns" them, recycling them into a new charge field that it can re-emit. Just as I have already shown how other quanta are engines in this way, the nucleus is just a bigger engine. A single proton, for instance, is already an engine, since it can re-process the charge field through these charge holes, feeding off the charge field and quantum field and then excreting the charge field by flinging it off via its rotation. This explains where the charge field comes from: we do not need to theorize that it is created from nothing by the proton, we simply allow that it is infinitely recycled. The same thing applies to the nucleus. The nucleus can also capture its own charge field through these charge minima and then re-emit it.

If this is true, then the various quantum beasts, including the periodic zoo of nuclei, are not only engines, they might almost be called alive, since they eat and excrete the charge field. Not only that, but they eat and excrete one another. The protons and nuclei aren't just passive valves through which pass the photons of the charge field. No, it appears that the protons and nuclei can eat electrons as well, digesting them by stripping off outer spins, and turning them into charge photons. Yes, a certain number of lucky electrons get caught in the whirlpool, and achieve a limited stability in the shells. But any electrons too high or too low for the whirlpool get sucked into the charge minima at the ends of the nucleus, and are turned into photons.

Our new model of the nucleus as a stack rather than a ball also helps us explain the relative paucity of elements. With the ball model, it is not clear why we cannot continue to add nucleons. Adding more marbles to a bag of marbles does not increase its instability or decrease its likelihood of existence. But with the stack model, it is clear that a larger stack implies a greater instability. For two reasons: 1) the stack becomes less probable the larger it gets, just as a matter of statistics. The neutrons have to arrive in the proper places at the proper times, in a fixed method, and this method becomes less probable with each added nucleon; 2) just as with a stack of plates, a stack of nucleons must become more unstable at greater numbers. To start with, the expanding length of the nucleus must encounter a larger cross section of the charge field, opening the larger nucleus up not only to photon wind variations, but to more quantum collisions of all kinds. As we know, a sphere is the most stable configuration of volume. A stack is among the least stable, at all levels of size. This must strongly prejudice nature in favor of the short stack.

THE MYSTERIOUS MUON

The muon is currently used as proof of time dilation, since without it, we are told, the muon would not live long enough to reach the surface of the Earth. The muon lives only 2.2 x 10^{-6} seconds, and travels only about 660m under normal circumstances. But it is thought to be produced at an altitude of 15km. It does reach detectors on the surface of the Earth, or at sea level, therefore it must be experiencing time dilation.

The problem is that physicists make this claim without applying the transforms directly to the muons they detect. Since they do not know the altitude of creation, they cannot know the distance or time. Transforms have to be done on data: that is, on measurements that are directly made in experiment. But we only detect the muons, or, in some cases, measure their energy. We cannot and do not measure time or distance, therefore we cannot apply time or distance transforms. The altitude of production is inferred from the energy, as is admitted.

This is a major problem since you cannot apply SR transforms to an inference. You cannot find that a body is dilated without knowing its initial state. We do not know the altitude of production for the muons we detect, therefore we cannot apply transforms. The current theory is circular. We claim that if the muons are dilated as we think they are, then they must come from a certain altitude. And they are able to come from that altitude because they are dilated. The theory has no content and no possible proof. If it cannot be proved, it cannot be proof of anything.

In fact, according to my corrected equations of SR, a muon in approach would actually be time compressed, not time dilated. Only objects moving away from us are time dilated. Applying SR to the muon with the right math cannot give us time dilation. Therefore the phenomenon must be explained in some other manner.

This is not to say that I doubt the theory of muon creation. In fact, I can use muon creation and detection as proof of my own mechanics only because I trust the experiments. I believe in the creation and detection, as it is given to us in current journals and texts. I just don't believe in the given time dilation, since there can be no time dilation without a measurement of time or distance. You cannot claim to transform a raw detection, and the SR transform for approach should show time compression anyway.

Muon detection is easy to explain without time dilation, provided you know how to apply the ordinary gravitational field equations. Using gravity in an orthodox way, the muon arrives on the surface of the Earth because it is accelerating toward the Earth, not just moving at a constant velocity. This must affect the apparent time of the trip.

This simple statement is controversial because it appears to conflict with current theory in several ways. First, because the muon is in freefall, and is not in a gravitational curve, it doesn't appear that GR allows us to find a time differential due to gravity.

We can find a time differential due to SR, but not to GR. In fact, it is thought that GR prevents us from accelerating the muon at all: you cannot accelerate something that is already moving at .9996678c,* not even in a gravitational field. Or, you could accelerate it a bit, but you would need a much more powerful field and more time.

We can solve this apparent conundrum if we use Einstein's equivalence principle to reverse the vector of gravity, as I have done so many times before. The reversal makes the surface of the Earth move out during each dt, which makes the Earth move toward the muon during each dt. The muon's trip is shorter with each passing moment, which makes the time for the trip smaller. The muon doesn't have to go as far, therefore it gets there quicker.

Although many think this is illegal, it is done all the time, even by the mainstream. Richard Feynman does it in his *Lectures on Gravitation*, section 7.2, as I have shown. To calculate a gravitational blueshift, Feynman moves the bottom of his box up toward light that is coming down. Since he is basically repeating the math of the famous Pound-Rebka Experiment of 1959, the math for that experiment must do the same thing. I have shown that it does in a recent paper. The Earth must be given a velocity toward the light, or no blueshift can possibly be calculated.

You will say, "Even if all that is true, you still can't explain muons that way, since using the known acceleration of gravity, the surface of the Earth will only move 1.1×10^{-11} m in 2.2×10^{-6}s. Turning your gravity vector around doesn't explain anything." That would be true only in the case that we take the initial velocity of the surface of the Earth as zero. If the initial velocity is zero, then we can use the equation $s = \frac{1}{2} at^2$, which does give us the very small number above. But if the Earth does not have an initial velocity of zero, that equation won't work. Well, the surface of the Earth does have an acceleration of 9.8m/s^2, but it doesn't have an initial velocity of zero. Since these accelerations and velocities are relative to the center of the Earth, as I have shown, the velocity at the surface of the Earth during a given interval can hardly be zero. The only zero velocity in the field would be expected to be at the center of the Earth. The surface of the Earth is about 6,378km from the center, so it could not have a velocity of zero relative to that center. If it did, then its acceleration would also be zero. So let us see if we can develop a velocity just from the numbers we already have. If we use the above equation, $s = ½ at^2$, and solve for t using 9.8 and 6,378, we get $t = 1141$. In a field of acceleration of 9.8, the surface of the Earth is 1141s away from the center. From that we can develop the velocity: $v = 6{,}378{,}000/1141 = 5{,}590\text{m/s}$. That is just a rough estimate, since the Earth's field is not 9.8 at all distances; but it gives you the idea of how things work in a gravity field.

My critic will say, "Even that won't help you, because at that velocity, the Earth still goes less than a meter in the given time." True, but the Earth is not moving at a constant velocity, it is accelerating from that speed. When we have the muon moving at the Earth, we do not have two velocities being added, we have a velocity of c plus an acceleration of *g*.

Now to solve. I developed a velocity for the surface of the Earth to prove that it could be done, and to show that if we reverse the gravity vector, the surface must have an initial velocity. But I don't wish to use that velocity in my solution. Since I have already shown that the vector situation allows us to move the Earth toward the muon, even while the muon is at c, let us do the math in the simplest way possible. Let us dispense with the vector reversal and keep the Earth steady. This will have the mathematical and mechanical result of giving all our motions to the muon, but it will not contradict all I have shown above. Doing the math this way is a convenience, and implies nothing about the physics involved. Although this will take the total calculated velocity of the muon alone over c, as a matter of addition, it won't concern us. Again it won't concern us, and won't break any of Einstein's rules, because due to the vector situation, the velocities aren't really added, they are subtracted. We apply them all to the muon only to simplify our math, but this in no way implies that the muon is actually going over c.

Everyone else who has looked at this problem has tried to apply the old equation $v = v_0 + at$, which fails to help us here. Even before Einstein forbid us from exceeding c, this equation couldn't solve problems like the muon problem, where particles approaching the Earth at c were supposed to be accelerated by *g*. It doesn't solve our current problem because it only gives us a tiny correction, one that is frankly counter-intuitive. A particle going c in a gravity field should be accelerated more than that, supposing that it can be accelerated at all. But rather than pull apart that basic equation [I have done that now, showing the equation is false in all cases], let us ask how long it would take to accelerate a particle in freefall from zero to c in the Earth's gravitational field. Let us pretend that the Earth's field is constant at all distances, and that we can allow bodies to freefall as long as we like. In that case, we lose the v_0 in the equation above, and it reduces down to $c = at$. Solving for t gives us $3.0591067142857 \times 10^7$ s. Since the acceleration is constant, the average velocity is c/2, and the distance traveled is therefore $4.58548560580009 \times 10^{15}$m.

So it takes this hypothetical particle almost a year to go from rest to c. Now, a nice question is, how far would it travel in the next second? In other words, how far would it travel if we continued accelerating it past c, at the same rate of 9.8? Easy, since we just add a second to the time and solve: $x = 2.989 \times 10^8$m. How far would it travel in the next 2.2×10^{-6} seconds?: $x = 617$m.

Hopefully, you see what I have done. Rather than assume that the muon is accelerated independently of its velocity, we assume that it is accelerated as if it were already in freefall. We assume the Earth cannot tell the difference between a body it has been accelerating for a long time and a body that just arrived or was just created. In other words, the Earth accelerates the velocity, not the body. In a gravitational field, the equation $v = v_0 + at$ doesn't work. In that equation, the acceleration is independent of the velocity. But we want to accelerate the velocity. To do that without calculus, you do it as I just did it.

My critic will say, "Very ingenious, since you doubled the distance traveled by the muon. It would have traveled 660m and you have added another 617m. This would double the "lifetime" of the muon as well, if we gave it to the time instead of the distance. Unfortunately for you, we don't require a doubling, we require a increase of 50x. Your muons are still 13,723m short of the surface of the Earth."

Sad that my critic still can't comprehend my numbers, this far into the paper. I didn't just find that the muon traveled another 617 meters on top of the 660; no, I found that the muon was traveling 617 meters while it was traveling 660 meters. Remember, it would travel 660 meters in no field at all. Then I found that a gravitational field would accelerate it 617 meters. I found, specifically, that the gravitational field would accelerate a particle that far in an interval of that length, assuming the particle never had an independent velocity of its own. Remember, we accelerated this particle from zero, so the particle had no velocity uncaused by the field. So if we want to find how far the muon can travel during that same interval, with both the acceleration of gravity and its own velocity of c, we can't just add the two numbers. We have to multiply them, giving us 660 x 617 = 407,220m.

If my math were complete, the muon could be over 27 times higher than it is currently proposed to be and still reach the Earth. But my math is not complete. I let the acceleration of the Earth be 9.8 over the entire trip of the muon, but the acceleration of the Earth is not 9.8 at an altitude of 268 miles. It varies over the trip, going from about 8.62 at the start to 9.8 at the end. I will not do the full math, I will simply estimate a new number. Our average acceleration will be 9.21, which gives us x = 660m, which increases our altitude to 435,600m, putting us way up into the ionosphere.

Take note that the two numbers match. We came into the problem knowing that the muon traveled 660m with no acceleration, then found it was accelerated 660 by the field of the Earth. Coincidence? No, of course not. We found that number precisely because the field is accelerating the initial velocity, not the muon. The initial velocity is not added at the end, as in the naive equation $v = v_0 + at$. The initial velocity is integrated into each and every differential of acceleration, as it must be. Which means the initial distance traveled during each differential is also integrated into the field acceleration. This gives us the equations

$$v_f = v_0 + 2v_0^2t$$
$$x = v_0^2t^2$$

And this allows us to estimate the distance traveled in the field by simply squaring the distance traveled outside the field.

$x_0 = v_0 t$
$x = x_0^2$

Those equation are rough, and some have not been satisfied by this verbal math that I like to do. So here are the full equations, so that you can see exactly why 660 comes up twice.

$c = at_0$
$d_0 = ct_0/2 = c^2/2a$
$d = v_f(t + t_0)/2 = a(t + t_0)/2$
$d = a[(c/a) + t]^2/2 = [(c^2/a) + 2ct + at^2]/2$
$\Delta d = d - d_0 = (c^2/2a) + (2ct/2) + (at^2/2) - (c^2/2a)$
$\Delta d = ct + (at^2/2)$

Because t is very small in this problem and c is very large, the second term is negligible. This makes the distance traveled during each interval due to the acceleration almost equal to the distance traveled due to c. This is why there is no acceleration variable in my equations above: it can be ignored when the time period is so small. A field of 9.8 will act pretty much like a field of 1 or a field of 100.

This finding will be of great use to physicists, supposing some few of them dig out long enough to recognize it. Their current number for altitude of muon creation is about 9 miles, which is way too low. They have kept it low purposely to blend more easily with this time dilation theory. The higher they create the muons, you see, the more dilation they need. They already have gamma at around 39, which is embarrassing enough. Any additional altitude will just make that number go higher. But they need the muon creation to be much higher than 9 miles, since as it is, they have muon production just above the troposphere, in the lower levels of the stratosphere. That doesn't make any sense.

My number, which comes out to be about 270 miles, is much better, since we are then in the upper levels of the ionosphere. In the ionosphere, we would expect muon creation. We need those ions for muon creation, and there just aren't enough of them at 9 miles to explain the number of muons we see.

Some will say, "That all works out pretty well, but didn't you say that the muon was time compressed? In another paper, you say that time compression is indeed equivalent to life extension, and this explains life extension in particle accelerators, where particles are approaching detectors. Shouldn't you have to calculate time compression here?"

No, I shouldn't *have* to calculate time compression here, since I have shown that neither time compression nor time dilation is necessary to explain the detection of the muon at sea level. Given the muon's known lifespan, I have been able to take its altitude up to 270 miles, with no discussion of Relativity. We can do a time transform if we like, and yes, it will show time compression and apparent life extension with the muon in approach. But there is no physical reason to do a time transform here, since 1) we don't require it to explain anything, and 2) we still aren't measuring the time or distance the muon has gone. In other words, I haven't devised an experiment above in which we are measuring the distance directly. We are not going up to 270 miles, creating muons, then measuring their arrival at sea level. I am just using gravity equations to show that the gravity field of the Earth can appear to accelerate particles that are already at c, and that it can do it without contradicting Relativity. It can do this because the gravity field acts physically or mechanically as a vector pointing out from the center. It acts to decrease the effective distance the muon must travel, and because it acts like this in the equations, it must act like a vector pointing out.

Finally, in closing, I will repeat what I have said in other papers: I agree with Relativity, for the most part. If we do direct measurements on a muon, we will not be able to measure it going over c. Likewise, if the muon measured us, it would not be able to measure a velocity over c. However, that fact does not impact this paper, since in calculating an altitude for the muon and in analyzing its trip, we aren't measuring its velocity. We are calculating a distance, which is not at all the same thing.

To say it one final time, in my equations, the muon is not going over c. The raw velocity is 660c, yes, but no real object is going that fast in any system, neither in its own nor anybody else's system. That number is a result, not a real velocity. It is a result of applying all the motions in the math to the muon, but that is just a convenience. The Earth's field also has a motion here, and integrating the two motions gives us a number over c. That is not forbidden by Relativity.

Conclusion:

A poor understanding of vector math caused physicists to believe that the muon and other incoming particles could not be accelerated by the Earth's gravitational field. They tried to add the motions, taking the total above c. But as a matter of vectors, the motions subtract. The distance between the approaching objects gets smaller, which means the velocities must subtract, not add. Once this is recognized, and the gravity field is fully understood, the muon can be accelerated without any conflict with Relativity or the limit of c. Furthermore, in the gravitational field (or any other field), the equation $v = v_0 + gt$ does not apply. Instead, when we are accelerating particles already at c, we must integrate the acceleration with c, in order to find a total distance traveled. You do not add freefall to the local velocity, you integrate the two. More generally, the common equation $v = v_0 + at$ does not work when a is a field acceleration. It only works when a is an internal acceleration, as with a car and its engine.

THE FINE STRUCTURE CONSTANT
and Planck's Constant

In his book *QED*, Richard Feynman has a final chapter called "Loose Ends" where he tells his audience some of the remaining unknowns of the theory of quantum electrodynamics. Chief among these is the number 1/137.03597, which is the fine structure constant. Feynman calls it the observed coupling constant or "the amplitude for a real electron to emit a real photon."[1] But at a place like Wikipedia, you will find it listed under "fine structure constant." Feynman says that "all good theoretical physicists put this number up on their wall and worry about it."

I don't worry about it because I know it is more misdirection. Feynman says that "a good theory would say that *e* is the square root of 3 over 2 pi squared, or something." But I have an even better theory: **The constant is a fake number**: an outcome of math specifically designed to keep you from looking in the right place.

The standard model defines the fine structure constant like this:

$$\alpha = e^2/2hc\varepsilon_0 = e^2 c\mu_0/2h = 2\pi ke^2/hc$$

Modern physics loves to bury mechanics under constants. As you can see, the fine structure constant, which is already a constant, is defined in terms of other constants, like the permittivity and permeability constants. Charge is also now buried under many other constants, including the Rydberg constant, the Josephson constant, Faraday's constant, Avogadro's constant, and more. Now, we don't want to have to fool with the permittivity constant or the vacuum permeability, since I have already shown that they are misdirections. So we will look at the third equation.

$$\alpha = 2\pi ke^2/hc$$

At first it is difficult to see what Feynman is asking. He asks why the number is 137, but in the first instance, it is 137 because of the way the equation is built. So why is the equation built this way? You can see that we have Coulomb's constant, but since we are dealing with quanta, we don't need it. I have shown that Coulomb's constant is a scaling constant, taking us from the quantum level to our level. But this equation isn't scaling anything to our level. Yes, light is going c relative to us, but it is also going c relative to the quantum level. Both the electron and photon are already at the quantum level, so to me the presence of k is a sure sign that these physicists don't have any idea what they are doing. That is how I know this fine structure constant is a ghost.

The only physically assignable variables or constants we have here are *e* and c, so Feynman must be asking why the relationship of c to the squared charge of the electron is what it is. Notice that the "coupling" is between a squared charge and a velocity. That's rather odd, wouldn't you say? For the coupling constant to be meaningful as a number, it should couple a mass and a mass, or an energy and an energy, or something like that. As it is, this number is just an outcome of a juggled equation, juggled purposely to hide the real interactions.

This fine structure equation, with h and k, is already too complex. But it was not complex enough for modern physicists, who were afraid some graduate student might unwind it. So in the decades since they have created even more complex equations, like this one:

$$h = \frac{M_u A_r(e) c_0 \alpha^2 \quad \sqrt{2} d^3_{220}}{R_\infty \quad\quad V_m(Si)}$$

Where R_∞ is Rydberg's constant,

$$R_\infty = \frac{m_e e^4}{8\varepsilon_0{}^2 h^3 c_0} = \frac{m_e c_0 \alpha^2}{2h}$$

Every decade, basic physics and mechanics is plowed under by more and more needless math.

Feynman's question should have been, what is the relationship of the electron's mass to its charge, or what is the relationship of the electron's energy to the photon's energy. He and his colleagues couldn't answer these questions because they had already buried them under so much math, but I can answer them quite easily. To do that, we first have to dig Planck's constant out of the rubble and show what *it* really is.

If we go to the Wikipedia page on Planck's constant and scroll down to the section called "origin of Planck's constant," we find that Planck himself had no idea of the value of the constant. He was working, like Newton before him, with proportions. In looking at Wien's displacement law, Planck proposed that the energy of the light was proportional to its frequency, and then simply made up the equality with his constant. In other words, he had no idea where the constant was coming from. Planck did not develop the equation from mechanics, he developed it from experiment: specifically, the experiments at the turn of the century on black body radiation.

That Planck had no idea where his constant was coming from is understandable, but that later physicists could not figure it out is beyond belief, especially after Einstein gave them the equation $E=mc^2$. Planck's constant is now taught as a conversion factor between phase (in cycles) and action. But action is an old feint: a longstanding blanket over mechanics. So we can ignore that. The constant is expressed in eV seconds, erg seconds, or Joule seconds, all of which are unhelpful mechanically, so we can ignore them as well.

I will now show that Planck's constant is very easy to derive mechanically, which makes it astonishing that the derivation is not on the Wiki page or in any textbooks. Once you see how easy it is, you will agree that this information must be hidden on purpose. There is no way that a century of particle physicists could have been ignorant of what I am about to prove, so we must assume they were hiding it with full intent to deceive.

We take Einstein's famous equation and apply it straight to the photon. We don't need the transform *gamma*: *gamma* applies to everything except light. Light is a special case, remember? Einstein's postulate 2? So we can apply the equation as is, with no transform.

$$E = mc^2$$
$$c = \lambda\nu$$
$$E = m(\lambda\nu)^2$$
$$h = m\lambda^2\nu$$

Now, take a common photon like the infrared photon, with a wavelength of about 8×10^{-6} m. In that case $\lambda^2\nu = 2{,}400$. So,

$$h = m(2{,}400)$$

Planck's constant is about 2,400 times the mass of the photon. You will say, "But the photon doesn't have mass!" And I say, that is what they want you to think, which is why they never use Einstein's equation on photons. Giving the photon mass, or even a strict mass equivalence, would bring down the entire structure of QED, so they can't let you go there.

You will then reply, "But your math is just circular. You haven't explained anything mechanically."

So I will. We start with the difference between the mass of the electron and the mass of the nucleon, which is called a Dalton, and which is about 1821. I have already shown that this number comes from the stacked spins on the electron, and I developed an equation that yields not only the Dalton but all the meson levels as well. In other words, I gave a mechanical explanation of the number 1821, with simple math and simple motions. I showed that this same quantum equation will give us the photon mass as well, by assuming the photon inhabits a fundamental level of the equation, just like the electron, nucleon, and all the mesons. This fundamental level is 1821^3 beneath the proton level, or 1821^2 beneath the electron level. All we have to do is multiply the proton mass by $1/1821^3$, which gives us:

$1.67 \times 10^{-27}(1/1821)^3 = 2.77 \times 10^{-37}$ kg

That is the mass of the photon, derived without Einstein's equation. So my math is not circular.

But is it the correct math? Let's see. If we multiply that mass by 2,400 we get 6.63×10^{-34} kg, which is, sure enough, Planck's constant.

I have proved my point. Planck's constant is hiding the mass of the photon.

But how does this answer Feynman's question? We have to go back to the fine structure constant and remove all the fudge.

$\alpha = 2\pi k e^2/hc$

I have shown in other papers that k and π are ghosts, and in this paper I have shown that h and α are ghosts, so we have to dump them. We will use their numerical value to absorb them into the equation.

$e^2 = hc\alpha/2\pi k = 2400mc(.0073)/5.65 \times 10^{10} = .091m$
$e = .3\sqrt{m}$
$e = 1.602 \times 10^{-19}$ C
$1C = 2 \times 10^{-7}$ kg/s
$\boldsymbol{e = 3.204 \times 10^{-26}}$ kg/s
$e = 6.08 \times 10^{-8}[\sqrt{kg})/s](\sqrt{m})$

So, Feynman's question becomes "How do we explain this numerical relationship of m to *e*?" Well, we can't do it from these equations, as you now see, since these equations are not giving us a number relation between m and *e*. They are giving us a number relation between m and e^2. To get the right dimensions for *e*, the dimensions for that last constant must be √kg)/s. Since there is not a 1-to-1 relationship between s and √kg, even that last number is not telling us what we want to know.

We have more work to do. Let's look first at that number for *e* in the next to the last equation, which is the current one. I have expressed it in kg/s, and this brings a lot of things to light. Remember that the electron has a mass of 9.11×10^{-31} kg. According to this equation the electron is emitting a charge every second that outweighs it by 35,000 times. The electron is emitting the mass equivalent of 35,000 electrons every second, or 1.16×10^{11} photons per second. So it is not just my charge field that has mass. The standard model charge field has a huge mass, it is just hidden by these dimensions like the Coulomb. Ask yourself why the standard model and textbooks never write the fundamental charge as kg/s. Textbooks tell you that charge is mediated by virtual photons, but they don't tell you that the electron emits 35,000 times its own mass of these virtual photons every second, just to create charge. You see, if they told you this, they would have to field your next question, which is, "How can the electron emit so much mass and not dissolve? How does this conserve energy?" In my theory, I put that question out in the open and try to answer it, but the standard model prefers to dodge it with all their sloppy math and undefined constants and complex dimensions like the Coulomb and Ampere and Volt.

What allows us to solve this easily is the loss of the constant k. Remember that I said k is a scaling constant, and we don't need it here. The reason is because in these equations we are comparing quanta to each other: no scaling is involved. For the same reason, we can import a trick I used in my quantum gravity paper, where I showed that as long as we are staying at the quantum level, and not scaling, we can use a very familiar number for gravity at the quantum level. If we are measuring gravity at the quantum level from our level, then we have to scale down using the radius as the scaling transform. But if we are *not* scaling, we can use 9.8 m/s^2 as the number for gravity. I showed that if the quanta measure their own gravity, this is the number they would get. It sounds crazy, I know, but I will show how it works again here. We just find a unified field force for the proton, using its mass and its acceleration.

$$F = ma = (1.673 \times 10^{-27} \text{ kg})9.8 \text{ m/s}^2 = 1.639 \times 10^{-26} \text{ N}$$

Multiplying by two to represent the vector meeting of the fields of both electron and proton gives us 3.279×10^{-26} N. Amazingly close to our bolded number above for *e*, isn't it?

You will say "Yes, but you have a pretty significant difference, 7.5×10^{-28} N. You also have the wrong dimensions. The elementary charge is in kg/s, and your number is in N."

Let's look at my margin of error, first. If we divide, we find my error is about 2.3%. But I have already shown in my papers on the Bohr magneton and Millikan's oil drop experiment that the Earth's charge field is skewing all experiments done on the Earth. It is responsible for the .1% difference between the Bohr magneton and the magnetic moment of the electron. It was responsible for Millikan's error. And so it must also be responsible for a .1% error in computing quantum masses. The proton's mass is determined in experiments done here on the Earth, and physicists have never included the effect of the Earth's charge field, since they don't know it exists.

You will say, "Your error is 2%, not .1%". First of all, it is not my error: it is the standard model's error. And the error enters this problem in multiple places. Just as in Millikan's oil drop experiment, we have a confluence of errors. Let's look at the mass spectrometer, used to "weigh" the proton:

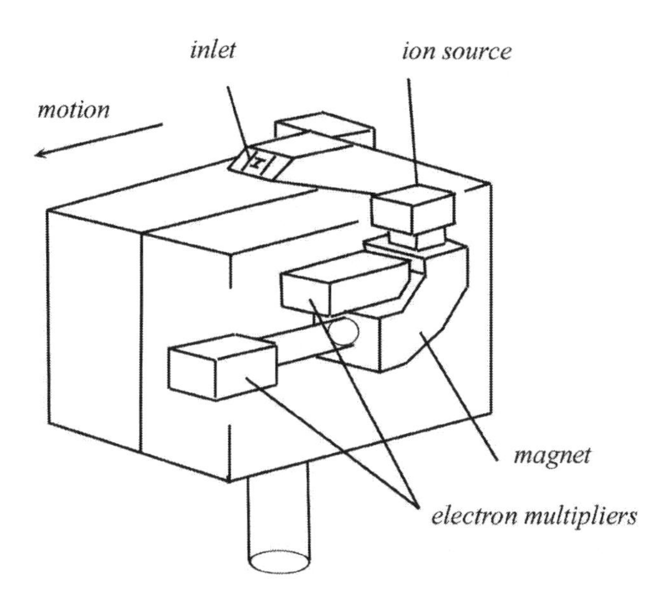

As you can see, the spectrometer must suffer the same problems as the oil drop experiment, since the magnet is in the plane of the Earth's charge field. The ions are moving straight down to start with and have a downward vector throughout the experiment. This can't work. The magnetic field is also rather weak, so it has no chance of burying the error simply by field strength.

But even if the machine is turned 90°, so that all motion is horizontal instead of vertical, the problem will remain. Unlike Venus, the Earth is both electrical and magnetic. If the experiment is done vertically, the electrical field of the Earth interacts. If the experiment is done horizontally, the magnetic field interacts. Both fields have the same strength, as produced by the charge field, so you are damned either way.

Although the mass spectrometer, either horizontal or vertical, must encounter the Earth's charge field, it does not encounter it precisely like the oil drop experiment did. Millikan set up the his electrical field in vector opposition to the gravity field, and included gravity in his calculations. But the math of the mass spectrometer attempts to ignore gravity, as an experimental constant. Masses in mass spectrometers are not calculated from gravity (in the experiment), they are calculated relative to each other. Wikipedia admits that "there is no direct method for measuring the mass of the electron at rest,"[2] and this is also true of the proton. You can see that the proton must be moving in the spectrometer, and its path must be bent by a field. The relative bend then tells us the mass.

At any rate, gravity is present throughout the experiment, and though it can be ignored as a matter of relative mass, it cannot be ignored mechanically. Because it is present, it must be included in any correction. Both it and the induced magnetic field are affected, but because they are not in vector opposition we don't treat them the same as we did with Millikan. With Millikan, we applied the charge field correction directly to his electrical field, since he aligned them. Here we halve the correction and then take the square root to square the effect. We halve the correction because the motion of the particle in the curve goes from (nearly) all gravity to (nearly) all induced magnetic field. Look at the curve in the diagram. At the end of the path, the particle is not moving down at all. So we go from "gravity is the entire cause of motion" to "gravity is almost no cause of the motion." If we sum that path, from all to none, all being 1 and none being 0, then the average will be about ½, given a smooth curve. So we only get half our error during the experiment. We only get half of it, but we still have to take the square root, since the error affects both the gravitational field and the induced magnetic field. Two effects will give us an increased total effect.

The charge field of the Earth is $.009545 \text{m/s}^2$, which is .0974% of gravity. Half that is .000487, and the square root is .0221 or 2.21%. Above, my error was 2.3%, so I am now within .0009.

The rest of that error is probably due to my math alone, since, as a theoretician, I almost never carry my calculations past the thousandths place. I will let those who love precision fine tune my math.

Now let's look at the dimensions. I have a force; the standard model Coulomb reduces to kg/s or Ns/m. But remember that the standard model is not too picky about its dimensions. The cgs system is still used, and in that system charge was kg or Ns^2/m. Yes, before SI, charge used to reduce to mass, although they never promoted that fact. So the dimension changes with the system. It changes again with my system, so that charge is a force, not a mass. I can change the dimensions without changing the number, because s/m reduces to one in my mechanics. Charge is the mass of the photon field, but a mass cannot give us a strength of interaction or a force by itself. You need a mass and a velocity, as I have shown elsewhere. This will give you a field strength, which will give you a force. Well, velocity is m/s. If you multiply s/m by m/s, you get one, and the field dimension reduces to N.

Conclusion:

The elementary charge is not a charge, it is a unified field force. The standard model believes that forces at the planetary or astral level are all gravitational and at the quantum level are all E/M, but this is false. The forces at all levels are unified field forces. The elementary charge includes gravity. For this reason we can use Newton's equations at the quantum level. Newton's equation is a unified field equation, and if we use it correctly, we can use it at any level. The measured masses of quanta are unified field numbers. *All* masses are unified field numbers, since they represent compound motions and forces. Quantum masses are hiding *both* fields, and this allows us to calculate "charges" straight from masses, without Coulomb's equation and without Planck's constant.

The elementary charge is not only a unified field force, it is a compound of emission by both the electron and the proton. Even when we are measuring the charge of the electron alone, the field will be composed of proton and electron emission. You cannot study electron charge alone, or proton charge alone, since you cannot go anywhere in the universe where the charge field is unipolar. Even on the surface of the proton or electron, you will find a bi-polar field. The charge field is everywhere, and it its strength is everywhere determined by compound emissions.

[1] *QED*, p. 129.

[2] http://en.wikipedia.org/wiki/Planck%27s_constant

THE COMPTON EFFECT, DUALITY and the KLEIN-NISHINA FORMULA

The Compton Effect is an inelastic scattering of high-energy photons by electrons. It was observed by Arthur Compton in 1923 in an experiment with X-rays. Like the photo-electric effect, it has correctly been interpreted as proof of the photon theory of light. That is, it is proof of the "particle" half of duality, and of Newton's corpuscular theory. This effect led directly to the Copenhagen interpretation of light, where Bohr insisted that light was both particle and wave, but also insisted this duality could not be understood mechanically or logically.

To show how he reached this flawed conclusion, we may return to a historical sidestreet called the BKS theory. This is the Bohr-Kramers-Slater theory, compiled mainly by Slater to convince Bohr the Compton effect could be explained rationally. Slater put it together in 1924, but Bohr had already dismissed it within the year. In short, the theory was a dual theory, with photons being emitted as particles, then guided by a classical E/M field of spherical waves. This field was pre-existing, created by matter, and containing all frequencies. The mediator of the field was not the motion of electrons or photons, but was a virtual field of virtual oscillators.

You can see why the theory was ditched, since that last part about virtual oscillators is a dodge. Bohr understood that if you are going to try to be mechanical, you have to show some convincing mechanics. If you can't show some convincing mechanics, you might as well dodge all mechanics from the beginning, staying with the math. He had learned this from Maxwell. Maxwell had done the same thing 60 years earlier. In the 1860's, Maxwell had tried to create vortices to explain the field mechanics, but, finding himself under heavy fire from Kelvin and others, he decided to give it up and go to fancy maths like quaternions instead. As you can see, Bohr and his pals did the same thing in the 1920's. They tried a few half-hearted efforts at mechanics and then gave it up for fancy maths.

But what is most interesting is that although we are told the BKS theory was ditched, we find that it is *still* the only existing pseudo-mechanics of the charge field. Now, almost 90 years later, we have no better explanation of the charge field or mechanism of quantum interaction. In fact, we have the *same* explanation: the field is now said to be virtual. Having nothing else, the mainstream has decided to embrace the worst part of Slater's theory.

To be fair, it is only alternative theories that now call this virtual field a field of oscillators. Thinking themselves revolutionary, many alternative theorists now propose some sort of mystical and undefined oscillators as the foundation of the charge field and the quantum field. They may use fancier shapes for their oscillators than Maxwell or Slater used, but their theories suffer from the same squishiness Maxwell's suffered from 150 years ago: the vortices, oscillations, shapes, or other motions do not really explain the data in a rigorous manner. On the other hand, mainstream theory is not even as mechanical as these alternative theories: the current standard-model explanation is a virtual field of virtual photons, these photons acting not by oscillation but by messenging or texting.

In either case, these virtual theories stop well short of a sensible explanation. No theory that contains the word virtual is sensible, since a virtual field is a ham-handed dodge of physics itself. If you are going to have virtual physics, you might as well have no physics, and at least Bohr understood that. He thought that if you can't explain something with mechanics, you had best dodge the issue with high-sounding philosophy or authoritarian fiats, and cover it over with as much math as you own. That is always preferable than being caught talking about virtual fields, while calling yourself a physicist or claiming to do physics.

Although current theory on Compton scattering and Thomson scattering is much more filled out than current theory on Rayleigh scattering, it is still very incomplete. It is incomplete because no one has been able to say how the duality expresses itself. It is well understood that Slater's field doesn't work, since the data can't be fit to photons carried that way by field oscillations. Slater was trying to explain interference and polarization and so on with the field, and the Compton effect with the photon, but when it came to explaining how the particle interacted with the field, he was at a loss. Particle physics is still at a loss, which is why they stick with the math and dodge all mechanics.

But I can explain the mechanics. The problem is that everyone from Newton and Huygens to Maxwell and Slater and Einstein has tried to express the wave as a field wave. But the wave of light is not a field wave. The wave belongs to each photon itself, and this is what solves the problem. All these theorists could not get out of the rut of thinking of light as an analogy to sound or other field waves. Because all the waves they had been taught in school had been field waves, they naturally thought light must be a field wave, too. So when it was proved by Einstein that light was not traveling via an ether, they were stumped. If there was no field, how could there be a wave? No one has gotten past that apparent dilemma.

The dilemma is a false one, though, since it never required a field to show a wave. Any spinning particle can show a wave, and you do not need a field to create spin. You need collisions, yes, and a field of particles to create these collisions. But once we attach the wave to the spin of the individual particle, we do not need a field to *express* it. In other words, the field *causes* the spins on the particles by collision, but the field does not *transmit* the wave or carry it. The wave is not a shape on a background of particles, as with sound. The wave is the spin of each particle, and is carried by each particle.

To clarify, I will give you an example. In a football stadium, you can find two different types of waves, both caused by people. In the stands, you will see a wave caused by groups of people holding their hands up and then dropping them. That is a field wave, like a sound wave. But in the aisles, you will find individual people creating waves just by walking. Their legs create wavelengths, since each step creates a gap. If you mapped a person walking by following the gap between the legs, you would get a wave. This second example is the analogy to a photon. A photon doesn't have legs, it has spin. If we map the spin over time, this spin will create a wave. The photon has a local wavelength that is determined by the radius of this spin, and that local wavelength is stretched out by its linear motion. That stretched-out wavelength is the one we see and measure.

This immediately explains many things. It explains why individual photons can carry a wavelength, as in the two-slit experiment. It explains superposition, since individual photons can stack spins. It explains simultaneous longitudinal and transverse waves, since, again, the photon can stack spins. The electron can also stack spins. And it explains the Compton effect because we now have a way to connect the photon to the field. The photon is not carried by the field; the photon *is* the field. The electromagnetic waves are carried by the photons, not the reverse. The charge field is a field of photons to start with, and the photons tracked by Compton devices and other devices are traveling in a field of other photons. There are photons in the field before the tracked photons are emitted, and that explains all the field mechanics in a direct way.

Particle physicists will say that this doesn't conserve energy, but it does, since quanta larger than photons can recycle photons. We already know that quanta can absorb or emit photons. What we have not understood is that they are absorbing and emitting constantly, and that Compton effects and other effects are just emissions and absorptions *above* this baseline recycling of the charge field. Even electrons in stable orbits are emitting the charge field. Every existing spinning quantum is recycling the charge field all the time.

This mechanism of spin also explains inverse Compton scattering, since we only have to turn our photons upside down to explain it. Spins are reversed just by a pole reversal. Photons can be spinning CW or CCW relative to electrons, and in one case the angular momenta will add in collision and in the other it will subtract. A subtraction will increase the wavelength of the photon, and an addition will decrease the wavelength.

The mainstream resists this simple spin interpretation for two reasons: 1) because they were not able to see that spins could stack, or how they should stack. I have shown the rules for stacking spins (chapter 7), and they are just gyroscopic rules, the same ones we have in the macro-world. These rules give us four stacked spins and five variances, which explain all the data that is now explained by quarks. 2) If we give the photon a radius and a mass equivalence, we will have to redo or throw out a lot of gauge math, including a lot of finessing like symmetry breaking that has taken the mainstream decades to pile up. This piling up has garnered a lot of Nobel prizes and such, and so no one wants to let it go.

But in chapter 8, I have already shown you a simple spin equation that unifies the proton and electron, as well as all the mesons. I have gone down the wish-list, answering all the embedded mysteries, as well as many that were not even known to exist. The best thing the standard model can do at this point in history is suck it in and move on. Any more foot dragging will just make them look worse than they already look.

Now, the specific math of the Compton effect is called the Klein-Nishina formula, and it is known to be a very good predictor of data. It overthrew the Thomson formula, which not only had the wrong radius for the electron, it also didn't properly incorporate Planck's constant and the quantization. Unfortunately, the Klein-Nishina formula, although much better as a heuristic equation, is still a mess. One way that it is a mess is in its use of the Compton radius for the electron. Although the Compton radius is now written in terms of the fine-structure constant, it is still the same *value* as the old classical electron radius used by Thomson. I have shown in chapter 10 that this value is too large by 100 times.

The current derivation dodges this by admitting that the Klein-Nishina formula "may also be expressed in terms of the classical electron radius $r_e = \alpha r_c$, but that classical quantity is not particularly relevant in quantum electrodynamics. [Wiki]" What they don't admit is that by that equation, the Compton radius is actually 137 times larger than the classical radius, making the Compton radius about 3.9×10^{-13}m. It is unclear how giving the electron a radius that large is "relevant" to QED, since it implies that the electron is only 100 times smaller than the Bohr radius itself. Basically, the Klein-Nishina formula matches data by giving the electron a radius larger than the proton.

So we need to rewrite the Klein-Nishina formula in a more logical way, keeping its result the same. Since I have shown the effective diameter of the electron is more like 9×10^{-17}m, that gives me a bit of work to do. [We use the electron diameter, not the radius, since the whole width of the electron is scattering; and we use the diameter with all spins, because the spins, having energy, also interact in scattering].

Let us see if we can make the correction simply by looking at the first term in the Klein-Nishina formula:

$$\alpha^2 r_c^2/2$$

The current numerical value of that term is about 4×10^{-30}. If we correct the radius of the electron, using my new value, we have

$$r_e^2 = 8.1 \times 10^{-33}$$

So we only need to raise that by a factor of about 490 to match current results. How can we do that? Well, let's study the other constant in the term. If we write the fine-structure constant in cgs, we have

$$\alpha = 2\pi e^2/hc = 1/137$$

If we express α in terms of h instead of h-bar (h-bar=h/2π), and thereby dump the 2π, then α has a value of 1/22 instead of 1/137. Since $22^2 \approx 490$, we have a match. This means we can rewrite the first term of the Klein-Nishina formula as

$$r_e^2/(e^2/hc)^2$$

Or, if we write α in terms of h instead of h-bar, then we can write that as

$$\alpha = e^2/hc$$
$$r_e^2/\alpha^2$$

The rewrite turned out to be fairly simple, as you see. Current theory had the wrong value for the electron radius, so they got the wrong value for the fine-structure constant, too. If we correct both of them, we can correct the Klein-Nishina formula without changing its result.

You will say that if we change the value of the fine-structure constant, we mess up other equations, but all the equations of quantum math are already messed up. I showed in the previous chapter that the fine-structure constant is a big fudge, so changing its value or its expression doesn't matter. We were going to have to ditch it anyway, no matter what we did here in the Klein-Nishina formula. The same goes for h-bar. Dirac's constant (h-bar) is written as an expression of angular frequency, but all the current angular equations are faulty. Because v ≠

$r\omega$, every angular equation ever used has to be rewritten [see chapter 3]. For that reason, all we really care about here is that $e^2/hc \approx 1/22$ in cgs. Whether we call that α or make up some new constant is immaterial. The important thing is not the constant; the important thing is that we correct the radius of the electron. Up to now, the fine-structure constant has only prevented us from doing that.

This is one way to correct the equation, but another is to recognize that the Klein-Nishina formula uses the ratio of photon energy before and after the scattering [$P(E_\gamma , \theta)$]. Therefore the first term we have been looking at could also be written that way. We notice that the term has a constant value (4×10^{-30}) very near the mass of the electron (9.1×10^{-31}). We only need to multiply the mass of the electron by about 4.4 to achieve the value of the term. If we assume the term is expressing an energy change in the electron to go with the energy change in the photon, then we could write the term this way:

$$E = ½m_e(c/v)^2 = 4 \times 10^{-30}$$

That gives us a ratio of the speeds of the photon and electron, so that what the term is telling us is that the electron is going about 1/3rd the speed of the photon in this experiment. But by writing the term this way, we also get a velocity in the equation, so that we can see how a variance in the electron velocity affects the scattering.

Some readers will find both those solutions tenuous, but I can show they are correct by setting them equal to each other and solving more problems.

$\frac{1}{2}m_e(c/v)^2 = r_e^2/(e^2/hc)^2$
$m_e/v^2 = 2h^2r_e^2/e^4$
$h = E/f$
$m_e/v^2 = 2E^2r_e^2/f^2e^4$
$E_e = \sqrt{(m_e/2)}fe^2/vr_e$

This gives us a way to calculate the energy of the electron without Relativity, by finding the frequency of the particle. Currently the frequency and wavelength are not known to change with an increase in electron velocity, but they must. The current equation for the Compton wavelength is

$\lambda = h/mc$

Which is a constant for any given quantum. For the electron this equals 2.4 x 10^{-12}m. But the wavelength of the electron should be dependent on its velocity. Current physics hides this obvious fact because they have no way of calculating this dependence. They hide it under the Relativity transform

$K_e = (\gamma - 1)m_ec^2$

I have shown that *gamma* is false in that and every other transform, but that last equation also hides the fact that the electron must be increasing energy due to increasing velocity and increasing frequency. I have shown that although Special Relativity is true, it is misused in cases like this to cover energy increases due to other causes. Just as Relativity has been misused in the atmospheric muon problem and the gravitational blueshift and a thousand other problems, it is misused here. We are told that the electron mass increases 100,000 times in an accelerator, all due to Relativity. That is false, and the falsity is all due to that false equation.

Logically, it cannot be just the relativistic mass that increases in an accelerator. The kinetic energy of the electron is increasing due to increasing speed and increasing energy input from the field. A large part of this energy can go into increased frequency, so we do not have to give it all to mass. But the last equation above hides this, because it has neither a velocity variable nor a frequency variable. We have no way of knowing how much the spin energy of the electron is increasing, since that equation seems to imply that all the new energy is going into mass.

If we use my new equation here, we can calculate an approximate electron frequency in the accelerator. If the maximum energy is about 50 GeV, and we assume the velocity is almost c, then the frequency is about 1.2×10^{37}/s. And the wavelength is therefore 2.5×10^{-29}m. That's the *local* wavelength of a high-energy photon, so we may assume that the accelerator has turned our electron into a photon, by stripping it of outer spins.

You will say that wavelength is way below the radius of my electron, which disqualifies my spin explanation. But, again, we have no indication that the electron at the end of such acceleration is still an electron. All we currently measure is the final energy of the particle. It is my belief that the accelerator has stripped the electron of its outer spins, so that it is no longer strictly an electron. Without its full complement of spins, the particle is now a photon. I have unified all the quanta, including the proton, electron, mesons, and photon, the only difference being the number of stacked spins. So it is quite easy to strip an electron down to a photon, by removing these spins. I have even done the math, showing the electron is 1821 times smaller than the nucleon, and that the charge photon is 1821^2 times smaller than the electron. In other words, the electron is 4 spin levels below the proton, and 8 spin levels above the photon.

For this reason, I believe that Relativity has prevented us from understanding what is really going on in accelerators. My explanation here is incomplete, but it is a step in the right direction. We must recognize that even at non-relativistic speeds, the wavelength of the electron must be dependent on its speed. Therefore the Compton wavelength cannot be correct.

The Compton wavelength, as currently derived, is not analogous to photon wavelengths, since when we measure photon wavelengths, we are measuring them at the macro-level: *as how we see them*. But I have shown that the local wavelength of the photon is much, much smaller, being on the order of 10^{-23}m (for infrared light). Therefore, the Compton wavelength of the electron must be a local wavelength of the electron, or the attempt at such. Since the local wavelength is just a particle radius, the Compton wavelength is the attempt to calculate the electron radius from Planck's constant. But, as I have shown, it fails in this, since the electron radius cannot be anything like that large, even if we include all the spins. The Compton wavelength is off by a factor of almost 10^5. The local wavelength of the electron is about 10^{-16}m, and the wavelength we would "see" would be stretched out by v^2. This is what I mean when I say that the electron wavelength is dependent on its speed. If we could accelerate the electron to c while the electron kept all its mass and spins, its "seen" wavelength would be something like 1m. Since we can't, we can instead calculate the "seen" wavelength of the electron at .25c: about 5cm. Therefore, we would expect an electrical field created by electrons moving that fast to either interfere with or augment microwaves of that wavelength, depending on the direction. Current physicists know that fields affect one another like this, but they aren't able to predict which fields will affect which, and to show why this affect is dependent on velocity. My new equations here will help them do that.

Conclusion:

I have shown that my new radius for the electron fits the Klein-Nishina formula like a hand in a glove. If my new number hadn't been very close to correct, we would not have been able to simply move the square constant from the numerator to the denominator.

THE DISPROOF OF ASYMPTOTIC FREEDOM *and the breaking of the Landau Pole*

Editor's Note: Some readers will find this chapter too mathematical and technical, but it is included here because it contains a very juicy bit of criticism near the end. Mathis tears apart the math of a recent Nobel Prize winner, using math from the physicist's own Nobel Lecture, and is able to make this transparent to us all. You don't see this sort of thing, well, ever.

Most people (even most physicists) don't know what asymptotic freedom is, but since three men—David Gross, David Politzer and Frank Wilczek—won the Nobel Prize for it in 2004, it is best we give it a look.

Asymptotic freedom is said to have revivified or rehabilitated particle physics, bringing it back from an abyss it faced in 1973. That is the year these men did their work, although they waited 31 years for their prizes. This abyss was an abyss in quantum chromodynamics, or the field in physics that deals mostly with quark interactions inside the nucleus. The problem was with the strong force, in which quarks are thought to play their part.

The strong force was proposed to overcome the charge force on the positive protons. Protons exist in the nucleus at very close quarters, despite having a strong repulsion. Therefore it was proposed that an opposite force overwhelmed the charge repulsion. This is the strong force.

The problem was, to make this strong force work, it had to change very rapidly. That is, it turned on only at nuclear distances, but turned off at the distance of the first orbiting electron. The strong force is an attraction, and we couldn't have it affecting electrons. Because the field had to change so rapidly (have such high flux), it had to get extremely strong at even smaller distances. Logically, if it got weaker so fast at greater distances, it had to get stronger very fast at smaller distances. In fact, according to the equations, it would approach infinity at the size of the quark. This didn't work in QCD, since the quarks needed their freedom. They could not be nearly infinitely bound, since this force would not agree with accelerator experiments. Quarks that were infinitely bound could not break up into mesons, for a start.

This problem existed for less than a decade before it was said to be solved. It was solved by proposing asymptotic freedom—which is a short way of saying that the math was pushed. Here is how the math was pushed.

First, we take the strong force and its flux as given. We have no direct proof of this field—it is only a postulate—but we assume that our assumptions are correct. In order to calculate the flux, we must calculate how the energy of the field approaches the upper limit. This upper limit then becomes an *asymptote*. You may remember from high school math that an asymptote is normally a line on a graph that represents the limit of a curve. Calculating the approach to this limit can be done in any number of ways, but Gross, et. al., did it by mirroring the math of quantum **electro**dynamics. QED had met this same problem, on a lesser scale (as I will show below). Lev Landau developed a famous equation to find what is now called a Landau Pole, which is the energy at which the force (the coupling constant) becomes infinite. Landau found this pole or limit or asymptote by subtracting the bare electric charge e from the renormalized or effective electric charge e_R:

$$1/e_R^2 - 1/e^2 = (N/6\pi^2)\ln(\Lambda/m_R)$$

I won't bother you with the right side of this equation yet, since the largest problem is on the left side. What we have here is a value for e obtained by one sort of math, and then another value for e that has been pushed by another sort of math to match the experimental value. We subtract one from the other to find a momentum over a mass (which is of course a velocity). [Take note that although momentum is normally represented by "p", we are told that Λ is a momentum in this equation.] Now, if we hold the renormalized variable e_R steady, we can discover where the bare charge becomes singular. Landau interpreted this to mean that the coupling constant had gone to infinity at that value, and called that energy the Landau pole.

I could begin my critique of all this by reminding my reader that renormalization is heuristics. Even Richard Feynman, the master of renormalization and inventor of much of it, admitted that, calling it hocus-pocus. The renormalized charge here is just a charge that has been pushed to match experiment. But even if we accept that renormalized math is genuine, one of our charge values here must be wrong. In any given experiment, the electron has one and only one charge, so that either e or e_R must be *incorrect*. Either the original math or the renormalized math must be wrong. If two maths give us two different values, both cannot be correct *in the same equation*. But Landau is telling us we can subtract an incorrect value from a correct value, to achieve real physical information!

Some will say that I have misunderstood the terms. They will say the bare charge *e* is not just an outcome of a variant math. They will say that the effective charge and the bare charge are *both* experimental values, of a sort, the effective charge being charge as seen from some distance and the bare charge being the charge on the point particle. In a way, the bare charge comes from 19th century experiments and the effective charge comes from 20th century experiments. The difference must then tell us something about the field. I realize this is the current interpretation, but it is factually incorrect, as those who interpret it this way must know. The bare charge on the electron contains a negative infinite term, just as the bare mass of the electron has (is) an infinite term. To get a usable figure, *both* have to be renormalized. Feynman got his Nobel Prize for renormalizing the *bare* mass, and for the bare charge to be used in an equation, it too has to be renormalized. Landau cannot plug infinities into his equation. Notice that the m variable also has an "R" subscript: that stands for "renormalized." So the fact is, both the bare charge and the effective charge are renormalized. Otherwise the charges would be infinite or undefined to start with. This means that neither charge is "an outcome of experiment." Both are an outcome of math, math that is not defined itself. Therefore it is absurd to claim that you can subtract one renormalized number from another, and achieve a meaningful velocity or a meaningful limit.

Renormalization is a trick, Landau's math is a trick, and Gross' math is a trick. So we have a triple-decker fudge here, nothing less. In reality, math like this cannot tell us anything about a limit or a pole or a maximum energy. If you subtract an incorrect value for a charge from a correct value for that same charge, the only information you can get is information about your margin of error. You can tell how wrong one of your maths is. But you can't tell anything about the flux of any field. Landau's math is complete and utter bollocks, nothing less.

As more proof of this, look at the *form* of the equation: the left side has potential values between 0 and 1. We can see this by multiplying both sides by the smaller charge.

$$1 - e_R^2/e^2 = e_R^2 (N/6\pi^2)\ln(\Lambda/m_R)$$

If e_R^2 is the smaller charge, then e_R^2/e^2 cannot be greater than 1 or less than zero.

If N = 4, and we set the value of e_R^2 at 1, then the natural log of the velocity must have values between 0 and 14.8. With a natural log of 14.8, the velocity would have to have a numerical value of 2.68 million. Lower values for e_R^2 will raise the value of the natural log, and therefore the velocity. For instance, if we measure velocity in meters per second, the charge on the electron must be very much smaller than 1. It must be around 10^{-19}. This increases the natural log to around 10^{20}, making the velocity $e^{100000000000000000000}$.

Let us say the bare charge and effective charge diverge so that one is double the other. This makes the left side ½, which lowers the natural log to 10^{19}, which lowers the velocity to $e^{10000000000000000000}$. At the speed of light, the natural log is 19.5, which means the charge values must have converged to within 10^{-19} (which, remember, *is* the charge on the electron). To get any appreciable divergence would require the electron to travel billions of times over the speed of light. For this reason, the Landau pole is meaningless. Even if the equation were in the correct form, the limit on the speed of a particle means that the charge values cannot approach these limits. The Landau pole is way beyond the velocity limit of the electron.

Again, my critics will say that I have pulled this velocity out of my hat. Landau's equation has no velocity in it. He never assigns the maximum momentum Λ to the electron itself, therefore I cannot assign the velocity to the electron. But again, Landau and current theory are wrong. Landau has the electron represented on both sides of the equation, as charges on the left side and as mass on the right side. This means the momentum variable will automatically assign itself to the electron. Landau may mean to assign it to the field or to another entity, but his intentions mean nothing to the numbers. The way the equation is written, that momentum must attach to the electron, giving us a velocity by the equation p = mv, so that p/m must equal v. Since that velocity has a limit, the charges must have limits that Landau and the standard model have never seen.

Even some mathematical physicists, using the same tricks as Landau and Gross, have come to the conclusion that something is wrong with the Landau pole. In the late '90's, there was a well-known "Landau pole problem" that made the pages of several journals. In one of them, the physicists claimed that, "A detailed study of the relation between bare and renormalized quantities reveals that the Landau pole lies in a region of parameter space which is made inaccessible by spontaneous chiral symmetry breaking."[1] Yes, as I have shown with very simple math, the Landau pole *does* lie in a region of space which makes it inaccessible to its own variables, and this region of space is inaccessible to them due to the limit on c. These physicists, with their complex math of spontaneous chiral symmetry breaking, were not able to tease out the real problem here, but with many pages of dense equations they were able to tell something was not right. There are many things not right with Landau's equation, but it has nothing to do with parameter spaces or difficult math. It has to do with assignment of kinematic and dynamic variables. It has to do with logic. Physicists have gotten in over their heads with these maths, and they cannot spot the flaws in even the simplest equations. They cannot, since they no longer study basic math and motions. They are buried under fancy operators and renormalized fields.

Gaps between renormalized values cannot yield energy limits, but Gross took this math of Feynman and Landau as bedrock. He accepted the Landau pole as legitimate, and using this math he calculated a Landau pole for QCD. This pole was way too high, so he needed a fix. He needed to lower that pole by a large margin. How did he do that? Well, with a lot more fudgy equations, of course. But under the equations lay the idea of **anti-screening**, which is the faux-mechanical explanation of asymptotic freedom. In short, Gross used the Dirac sea of virtual particle pairs (or the Higgs field) to explain the drop in energy at very close quarters in the nucleus. First, Gross inserted the vacuum in between quarks. This is legitimate (even to me) since no distance is infinitely small. You can always insert the vacuum. Then he proposed that the vacuum is made up of virtual particles. In the narrow confines of QCD, only the gluons can exist in this squashed vacuum, and they exist as another sort of virtual particle pair. In this case, the gluon is a color, anti-color pair. If the quark approaches too close to this color-anti-color boundary, the gluon faces it with a similar color, driving it away. This "screens" part of the strong force, explaining why it doesn't continue to increase at the smallest distances.

The idea of color confuses this analysis somewhat, and it is easier to understand screening by looking at the screening in

QED. In QED the vacuum is composed of positron-electron virtual pairs. As a real electron interacts with the vacuum, the virtual pair shows its positive face, attracting the electron even more. So in QED, this mechanism is used to explain the opposite phenomenon. In QED, we have screening, and in QCD we have anti-screening. In QED, the electron is attracted to the vacuum itself. This is said to solve other problems I don't have time to address here.[2]

What is wrong with this explanation? Many things. We start with the fact that neither gluons nor quarks have ever been seen, or tracked singly in accelerators. Neither have virtual particles. Nor has color. Color has never even been *defined* mechanically. Is it charge, is it spin, is it emission? No one knows. We don't even get the beginning of a mechanics. Just pages full of paragraphs like those above, proposing fields and particles and colors from nothing.

We also have a bald contradiction and *reductio* in this postulate that the vacuum is a sea of particles. It is a contradiction in that the word vacuum has always been used to mean "that thing that is not material or particulate." The vacuum is supposed to be nothing, by definition, but here it becomes something. That is a contradiction. It is a *reductio* because it begs the paradox of Parmenides. If the vacuum is composed of virtual particle pairs, then it is no longer the vacuum: it is matter. If everything is matter, then you have a plenum in which motion is impossible. Calling this matter "virtual" is just a dodge. The vacuum then becomes nothing when you need it to be transparent and something when you need it to have physical characteristics (like polarity). Defining something as both x and non-x is not physics, it is magic, sophistry, or pettifogging. No matter how many high-profile sophists or magicians we have (Dirac, Higgs, Yang-Mills, etc.), we cannot outrun illogic. You cannot turn a contradiction into a postulate, no matter how many authorities say you can.

But the basic problem is that we have an equally clear *reductio* here in the regression of fields. First, in order to explain a force, physics created a field called E/M. It never explained mechanically how that field worked, it just created field lines. To this day, we have no mechanical explanation of charge. Is it emission, is it spin, is it motion? If so, how does emission or spin or motion create both attraction and repulsion? No answer, either at the macro-level or the quantum level. Then, we look closely at this field and discover it doesn't work in the nucleus. Our opposite charges should create a repulsion. So we create another field that only works at this level, to counter E/M. But we don't define this field in terms of spin or motion or emission either. We only define it mathematically, and with cutesy names. Then we look closer, and we see that this field also doesn't work, in and around baryons. So we create another field. That is what the gluon field is. This anti-screening gluon field reverses the strong force just like the strong force reversed the E/M field.

But to be consistent, we must then look at the flux of the anti-screening gluon field. The standard model is sick of this stuff, so it just decides to go virtual at this point, dodging all further questions. But if the flux of the strong force was a problem, the flux of the anti-screening gluon field must be a similar problem. Gluons switch from attractive to repulsive over an even smaller area, so they must have greater flux. Gluons are called gluons because they are the "glue" of the hadrons, and glue is an attraction. But in anti-screening, they switch to repulsive. Therefore, we have a repulsive field underneath the strong force, and this force must have greater flux than the attractive strong force. In which case we need a fix for that also, so that *it* does not have a Landau pole creating an infinite coupling constant.

QCD answers this by throwing up its hands and saying, "We are beneath the Planck length now, and we refuse to answer any more questions!" But the problem is not one of size, it is one of logic. QED and QCD keep fixing problems in existing fields and particles by proposing sub-fields and sub-particles. But they try to do this by dodging mechanics. Instead of fixing the mechanical hole in the E/M field, they drive around that hole and build another sub-field. That is, instead of showing how charge is created mechanically, they propose the strong force, to counteract charge. Instead of showing how the strong force is created mechanically, they propose anti-screening, to counteract it.

But I have shown that if you build the E/M field with the right mechanics, you don't need the strong force [see chapter 19]. And if you have no strong force, you don't need asymptotic freedom to fix it. If you get your first field right, you don't need an infinite regression of sub-fields to fix your errors. We don't need all this illogical math and theory, since we can fix the E/M field with simple and logical postulates. We don't need to renormalize our equations: we need a theory that gives us normal equations to start with. With these normal equations and postulates, we don't need an infinite regression of repairs.

To be specific, many have interpreted asymptotic freedom as giving us a field like a rubber band,[3] which is slack at near distances and taut at greater distances. It is a sort of inverted field. But this begs the opposite question: "If the field is greater at greater distances, what causes the asymptote or limit at the other end? That is, why doesn't the strong force get stronger as you pass the nuclear shell? Why doesn't the strong force pull on electrons? Physicists have solved the problem by inverting it and then ignoring the inverted problem.

The answer to this question is that the strong force needed to be inverted in order to make it change like the E/M field is changing. The physicists think they are measuring strong forces between quarks, but they are actually measuring stacked spins on the baryon, and gravitational-E/M forces percolating through the spins.

They therefore have to invert the flux to make the strong field change like the E/M field. They have to make their theoretical field change like the real field, even though an attractive field cannot possibly increase with distance. There is no strong force, so it has to be reversed. Reversed, it magically has the same flux as the E/M field it inhabits, while having the opposite sign! This gives them a force field that acts non-mechanically and illogically, but it at least allows them to keep their strong force. But, as I hope you can see, it is much simpler to assign the reversed flux to a repulsive E/M field, which it fits. Then you can explain attractions as weaker E/M fields, instead of more powerful strong fields. You don't need finessed math to explain the inversion of the field, you just need a theoretical clean-up. An asymptotically free strong field mirrors the E/M field because it IS the E/M field. The strong force is just a subset of the Unified Field, and does not exist as a separate or separable field.

For instance, in 1964 Vanyashin and Teren'tev calculated the charge renormalization of vector mesons, getting the opposite sign they expected. The field flux was reversed, according to their math. They thought there was something wrong with the math. Quantum physicists now explain the sign with asymptotic freedom. But the real answer is that the vector mesons were not traveling in a field of "strong" vectors or potentials: they were traveling in a field of potentials created by gravity-E/M—a field of real B-photons that was mainly repulsive, but that was a compound of E/M and gravity. The field, though having the flux of E/M, appeared attractive because most of its strength had been turned off. It was relatively attractive, compared to the very strong E/M field outside the nucleus. And this is simply because most of the B-photons were emitted outside the nucleus, due to gyroscopic rules of spin.

As one final proof against asymptotic freedom, let us look at the math, such as it is. It should be a matter of interest that Gross had published, only a year earlier, and using very similar math, "a proof that no renormalizable field theory that consisted of theories with arbitrary Yukawa, scalar, or Abelian gauge interactions could be asymptotically free." [Coleman and Gross, 1973.] No one had shown this proof was wrong, but nonetheless Gross could see that the need in quantum physics for asymptotic freedom was greater than the need for proofs against it. QCD wanted asymptotic freedom, and Gross planned to supply it. He would change his course in any way required in order to supply it. If Abelian gauge theories were necessarily non-asymptotically free, he would pursue non-Abelian gauge theories. But all this talk of gauge theories is misdirection, as Gross proves in his Nobel Lecture, where he supplies "the arithmetic":

> The contribution to ε (in some units) from a particle of charge q is $-q^2/3$, arising from ordinary dielectric or (dia-magnetic) screening. If the particle has spin s (and thus a permanent dipole moment γs), it contributes $(\gamma s)^2$ to μ. Thus a spin-one gluon (with $\gamma = 2$, as in Yang-Mills theory) gives a contribution to μ of $\delta\mu = (-1/3 + 2^2)q^2 = 11/3q^2$; whereas a spin one-half quark contributes $\delta\mu = -[-1/3 + (21/2)^2]q^2 = -2/3q^2$ (the extra minus arises because quarks are fermions). In any case, the upshot is that as long as there are not too many quarks the anti-screening of the gluons wins out over the screening of the quarks.[4]

Gross then tacks on the formula for the beta function of the non-Abelian gauge theory, but that is just window dressing. You can see that he has already given us the math and the explanation, with simple arithmetic!

To begin with, notice the odd language here after the math. "In any case, the upshot is. . . ." I was struck by that the first time I read it. "In any case" is not applicable here, since Gross is not supposed to be giving us an example or a suggestion, he is giving us famous math. It is highly irregular to follow a mathematical proof with "in any case," as if all this is perhaps beside the point. "The upshot is" is also odd phrasing. We find nothing else like it in this lecture. As one would expect with a Nobel Lecture, this paper is not breezy and informal. It appears that Gross is subconsciously attempting to hurry us past this math, and trying not to put too much emphasis on it. Why?

Because he has just put two falsehoods in full view. He is afraid someone might notice this, but he can't help but do it anyway. Both falsehoods happen to reside in the sentence immediately preceding "In any case." The first falsehood is the last equation. The second falsehood is that the "minus sign arises because quarks are fermions." I would hurry past that, too, if I were Gross. The whole proof relies on it, and it is known historically that he and his colleagues changed the sign right at the end. At first they had the "wrong" sign, and then they changed it. This is a well-known part of the story, since Politzer claims to have gotten it right the first time (and claims special recognition for that). But we are dealing with spins and charges here, as you see. The final equations are $11/3q^2$ and $-2/3q^2$, and q is explicitly defined as charge. Well, according to the standard model in 1973 and now, gluons are spin 1, charge 0. Quarks are spin ½, and the charge may be either 2/3 or -1/3. The quarks in the proton and neutron (the most common quarks) are up and down quarks, which are 2/3 and -1/3, respectively. So the fact that quarks are fermions does not decide the question. Even according their own rules, these equations are misleading. That last equation already includes the sign of the quark inside the brackets. That first term inside the bracket can be either -1/3 or 2/3, which expresses the charge of the quark. There is no need or excuse for re-expressing the charge outside the brackets. Gross implies that all fermions are negatively charged, but even disregarding the three quarks with positive charges, we have the three neutrinos with no charge, according to the standard model. Even this simple arithmetic has been pushed!

This is crucial, since that minus sign decides, by itself, the asymptotic freedom of the field. It has to be the opposite sign of the gluons, so that the anti-screening of the gluons can counteract the screening of the quarks. If both signs are the same, we have no anti-screening and no freedom. Gross himself has proved my point, in his own Nobel Lecture.

You can't give quarks an extra minus sign just by categorizing them as fermions. "Fermions" is just a name. A minus sign is supposed to signify some mechanics. Since the charge has already been signified, this minus sign cannot be a charge sign. It is not a spin sign either, since quarks are spin 1/2, and we have no 1/2 there. The minus sign is just a mathematical fudge, which is precisely why Gross loses his cool at that very point, reverting to the psychological pointer "in any case."

As I have said before, physicists don't know when to shut up. It was a magnificent blunder for Gross to publish this simple math, since it put the fudge in high focus. It would have been much better for him to continue to hide behind the beta function and the gauge fields, which provided some cover. But his Prize made him overconfident. Like a criminal who has dug up the loot after three decades, Gross couldn't help bragging. He has unmasked himself.

As a closer, I will draw your attention to the fact that people are now being given Nobel Prizes in physics without doing any physics. As I have proved, what Gross, Wilczek and Politzer did was bad math, not physics. Physics is supposed to be "physical," which means material and mechanical. Non-mechanical theories and mathematics are not physical. Virtual particles are not physical: if they were, there would be no need to call them "virtual." The Landau pole is not physical, since it is found by applying a mathematical margin of error to a problem, and claiming to have developed a number that can be applied to a momentum. But a momentum cannot be derived from a margin of error. That is like saying that you can manufacture a leprechaun from a bag of leap-years. It is mathematical alchemy of the most ridiculous sort.

Not only did these guys fail to do any physics, they failed to do any real math. They did only ghost-physics and ghost-math. They created a problem with ugly math, defined it with uglier math, and solved it with even uglier math. Ironically, each math created more problems than it solved. Math and science are supposed to solve problems, but ghost-math and ghost physics do the opposite. Each ghost spawns at least two more ghosts. This is great for job-creation, but terrible for anyone who desires a meaningful or physical explanation.

The 2008 Nobel Prize for physics was awarded to three more physicists working on basically the same problem. Yoichiro Nambu, Makoto Kobayashi, and Toshihide Maskawa won for their work on spontaneously broken symmetry. In the chapter on QCD [next], I show that spontaneously broken symmetry is more awful theory and math, so the prize is once again being awarded to fake physicists. Since there are no quarks, giving prizes to quark physicists is like giving prizes to fairy spotters. Also take note that the Nobel physics committee can't seem to find any real physicists to honor. Although physics is an astonishingly broad field, it appears to be winding down. The Nobel Committee keeps returning to the same narrow sub-field, and we may assume that this is because that sub-field, QCD, has benefited from the most promotion in the past several decades. As with Hollywood, physics is now mainly PR, and the Nobel Prize is simply the capstone in the long public relations campaign, the Oscar of physics.

I suggest that the Nobel committee discontinue the prize for physics, and substitute prizes for alchemy, job-creation, and propaganda. As with the Peace Prize, the prize for physics is now a bald misnomer. Call them the War Prize and the Fudged Math Prize.

[1]Göckeler et. al., arXiv:hep-th/9712244v1
[2]Mathis, Miles. *Screening in QED.* http://milesmathis.com/screen.html
[3]http://pr.caltech.edu/periodicals/CaltechNews/articles/v38/asymptotic.html
[4] REVIEWS OF MODERN PHYSICS, VOLUME 77, JULY 2005 , p. 844.

A REWORKING OF QUANTUM CHROMODYNAMICS
and a dismissal of the quark

This chapter is only an overview and first step in a very large undertaking: overhauling QCD. I will not be able to list all the corrections here, of course. That would require a book. At this time, I only intend to critique QCD very broadly, and to very broadly show the method for fixing it.

My first task here—showing that QCD is a very bald heuristics—is not a difficult one. Anyone who looks at QCD for more than a few hours will discover that most of the words in QCD are mechanically unassigned. The founders of QCD embraced this fact from the beginning, making no effort to hide it. That is why they chose words that displayed this heuristic character very plainly, with whimsical characteristics that were flagrantly non-mechanical. And so we have quarks with color, strangeness, charm, flavor, and so on.* Now, 40 years later, we have much more data, but we have not seen a mechanical explanation of the numbers. We have not seen this explanation because no one demands it or desires it: all are satisfied with the math and the terms.

For example, here is a quote from Murray Gell-Mann himself:

In order to obtain such relations that we conjecture to be true, we use the method of abstraction from a Lagrangian field-theory model. In other words, we construct a mathematical theory of the strongly interacting particles, which may or may not have anything to do with reality, find suitable algebraic relations that hold in the model, postulate their validity, and then throw away the model. We may compare this process to a method sometimes employed in French cuisine: a piece of pheasant meat is cooked between two slices of veal, which are then discarded.[3]

That's very clever, but I don't see any physics there. Ask yourself why contemporary physics is so keen to avoid "reality". No matter whose book you are reading, whether it is Gell-Mann's or Feynman's or Pauli's or Penrose's or Hawking's, you see this same sort of sentence. "We are not concerned with reality, because we don't claim to know anything about reality; we only want to match the math to the experiment." They not only state this very strange idea out loud, they parade it as a bold piece of theorizing. But ask yourself what, precisely, is bold about it. Is it really boldness, or is it cowardice? To me, physicists avoiding reality is not a sign of great boldness or invention. It is a ploy, a clever misdirection in order to avoid troublesome questions. The physicist is acting more like an attorney here than a scientist. A scientist would naturally be interested in reality, since that is what science is about. Only a sophist posing as a scientist would think of trying to convince you that reality is of no importance.

Notice that Gell-Mann is not only fleeing reality, he is also fleeing the confines of his own math. He has created a mathematical model, in order to skirt any questions of reality, then he has thrown away the model, too. That is like explaining A with B, then throwing B away. What do you have at the end? You have neither A nor B. You have no tasty pheasant or fatty veal, you have a sandwich of air. That would seem to be non-productive and hunger-inducing, but it is just what the contemporary physicist wants. You see, a theory of air cannot be critiqued. Try critiquing a cloud: there is nothing to grab onto. Just as a good attorney is infinitely slippery, so is the contemporary physicist. He is not a pheasant or a baby cow, he is an eel. Gell-Mann has created a theory that looks like something, but that is really nothing.

Wikipedia brags that "there is a huge body of experimental evidence for QCD,"—and this brag is at the top of the page on QCD—but near the bottom of the page, we find the astonishing admission that quantitative tests of non-perturbative QCD are few in number. We even find an "unsolved problems" section, in which we find this bombshell:

The equations of QCD remain unsolved at energy scales relevant for describing atomic nuclei.

The writer pretends this is a problem of "confinement" [see below], but it is a much bigger problem than that. It would be analogous to admitting that Newton's gravitational equation "remains unsolved at energy scales relevant for describing planetary orbits."

So both the terms and the overall theory of QCD are non-mechanical. As another example, all physicists will admit that color is a "hidden" variable—that is, un-assignable to any known or theorized physical or mechanical property of the quark. But most of the other words (quark quantum numbers or "flavors") are equally un-assignable. When a quantum physicist says that quark spin is assignable, for instance, he only means that we have built a table for it and added to the matrix. He does not mean that the quark is spinning in any known way, expressed by that number.

Quarks have a spin of 1/2. What does that mean, mechanically? Does it mean the quark is spinning half as fast as some other fundamental particle, or that it is spinning with a halved amplitude or frequency, or a halved angular momentum? No one knows.[6] I don't believe that anyone (but me) cares. Spin, like the strangeness number (S) and the baryon number (B) and so on, is mechanically unassigned. Wikipedia puts it this way: "elementary particles are believed to be point-like and so they cannot rotate around themselves." Most or all of these variables are hidden to very large degrees, in a mechanical sense, and the quark numbers are mainly tweaked with each new discovery, to keep the tables symmetrical and internally consistent, according to the manufactured rules of the theory.

The last quote above from Wiki brings up another problem. If quarks are point-like and cannot spin in a physical sense, what do chirality and symmetry apply to at this level? We have quarks that are right-handed and left-handed in various ways, we are told, and some interactions preserve symmetry and others don't. But what is moving clockwise or counter-clockwise, expressing right or left "handedness" if not real spin? How can a point express chirality, and how can a point be either symmetrical or non-symmetrical? To speak of symmetry or non-symmetry logically requires heterogeneity. It is illogical to assign chirality or symmetry to a point.

I will remind you that quarks have never been seen in bubble chambers, ionization chambers, or any other experiments. Quarks, along with all their quantum numbers, are wholly theoretical. They go part of the way in explaining experiments; we have no better theory; so we continue to tweek what we have. It is said that all the various quarks have now been "discovered," but this does not mean they have tracks all to themselves. It means that mesons have been discovered that come close to energy predictions. For instance, the meson sometimes called the charmonium was the particle detected as "proof" of the charm quark. But the charm quark was never seen: only the proposed pair.

This is true of all quarks. No quark had been confirmed with a visible and singly assignable track. As with the bulk of string theory, the complete non-arrival of the fundamental objects has not deterred anyone. In fact, one might say that the invisibility of the guest of honor is a necessary feature of the party. We are told that quarks are unstable outside the nucleus, but this is just a bald assertion, with no theory to back it up. Notice that we are never told why the quark should be so unstable, or more unstable than the other very unstable exotics in the quantum zoo. For example, we have managed to produce proof of the W boson—at least enough proof to convince insiders—though the W boson only lives for 10^{-27} seconds. Why should the quark be less stable than that? Both particles are normally buried in the nucleus, and the W is a mediating particle. It would seem it should be more fleeting than the quarks it mediates, not less.

Wikipedia addresses this question in passing (since the standard model itself only addresses it in passing). Wiki admits, "There is no analytic proof that QCD should be confining," which means there is no reason that the quarks should be confined to the nucleus, even in high-energy collisions, but Wiki continues:

Intuitively, confinement is due to the force-carrying gluons having color charge. As any two electrically charged particles separate, the electric fields between them diminish quickly, allowing electrons to become unbound from nuclei. However, as two quarks separate, the gluon fields form narrow tubes (or strings) of color charge, which tend to bring the quarks together as though they were some kind of rubber band. . . . When two quarks become separated, as happens in particle accelerator collisions, at some point it is more energetically favorable for a new quark/anti-quark pair to spontaneously appear out of the vacuum, than to allow the quarks to separate further. As a result of this, when quarks are produced in particle accelerators, instead of seeing the individual quarks in detectors, scientists see "jets" of many color-neutral particles (mesons and baryons), clustered together. This process is called hadronization, fragmentation or string breaking, and is one of the least understood processes in particle physics.[1]

"Gluons carrying tubes of color charge" as the mechanism of confinement is so outlandish it is difficult to believe it is paraded in the open as a scientific explanation. Physicists no longer know when to keep quiet. Are these tubes like the tubes of paint in a painter's box: does the gluon also carry a palette and brushes and a little bottle of turpentine, for clean-up jobs? And as for the rest, how does anyone know it is more "energetically favorable" to create quark-anti-quark pairs than to allow quarks to separate further? That is not a scientific statement. If we don't know the energy of each event, how can we know which is more energetically favorable? And we can't know the energy of each event, since we are told one event—quark separation—is impossible. This physicist implies that it is feasible to measure the energy difference between an unknown event and an impossible event! By this logic, any seen event must be more "energetically favorable" than any unseen event, and we can propose that things happen as they do because the energy was favorable. Yes, the quarks just have the wrong *ch'i*.

The only fresh air here is the admission that hadronization is not well understood, although I don't know why they aren't able to make up something. The rest of the explanation is manufactured from nothing: why not manufacture a long-winded answer for hadronization, and pass it off as bedrock?

There is no analytical or theoretical reason that the quark should remain hidden. Conversely, there is every reason to think that the quark's non-discovery should be a very big problem for the standard model. In any other field, and at any other time in history, a theory with such a giant, obvious hole in it would have major problems. A theory of cats without a cat would not be a strong theory. A theory of unicorns without a unicorn would not be a strong theory. But, for some reason, a theory of quarks with no quark is one of the most famous and feted theories of the 20th century. I suggest to you that this is a measurement of the health of the field and of the independence of those that inhabit it. How can it be that physicists are allowed to theorize with no least amount of criticism? On the contrary, the method now appears to be that a high-level person in some prestigious physics department comes up with a theory, and everyone immediately goes to work seeking confirmation of it. No one goes to work trying to show that it is a pathetic mess of illogic and question begging, although in most cases that is precisely what it is.

The decades of collision experiments in QCD have been very useful. I do not wish to imply that no good work has been done. Even the given theory of chromodynamics may be said to be a useful first step, in that it catalogs all the interactions and provides tentative variable assignments. But as a piece of mechanics or dynamics, it is threadbare.

One might say that chromodynamics, like electrodynamics, has mainly suffered from a lack of patience. Like QED, QCD has seen its data far outstrip its theory. QM and QED were built on the back of a woefully incomplete E/M theory, as I have shown. If we stick to the standard model, we still have no idea how to explain the E/M field mechanically, at a foundational level. The theory of positive and negative charges, mediated by virtual or messenger photons, is a mechanical disaster. Despite this, QM and QED found it necessary to explain more and more complex interactions, at smaller sizes. Quantum physicists did this by pasting together more and more math, and by ignoring mechanics more and more the farther along this path they went. Early on, the big names gave up on theory as mechanics, and said so out loud. Bohr and Heisenberg, in the infamous Copenhagen Interpretation, announced that mechanics, as a logical explanation, was categorically impossible. From that point, we could hope to have math, but we could not hope to have mechanics. We could not hope for a logical, physical explanation. This attitude has remained up to the present moment, transferred through Pauli and Dirac and then through Feynman and Gell-Mann, and now personified by most living physicists in most current physics departments.

From its inception, QCD followed this program precisely, continuing to ignore mechanics and dynamics and kinematics. It continued to use these old words, since they carried prestige and provided a feeling of solidity. But, by the old definitions, there is no mechanics in quantum mechanics, and no dynamics in quantum chromodynamics. There are only the math and vocabulary.

QCD had to be built on the back on QED, but the back of QED was broken to start with. QED's back was broken because it inherited the fields of Faraday and Maxwell, and never clarified them. E/M has no foundation, which gives QED no foundation, which gives QCD no foundation. And we have now built our structure to such a height—we are so far above the ground—that we may be said to be floating upon thin air.

Ironically, QCD has suffered from two opposite syndromes. I have just shown how the data outstripped the theory: QCD had to be built on the back of very incomplete E/M, QM, and QED theories. But the theory also outstripped the data. It did this by once again rushing. QCD was not patient enough to fill in the gaps in QM and QED before it began theorizing in its own arena; and it was not patient enough to wait for data from accelerators before it began theorizing fundamental particles and interactions. Gell-Mann and Zweig first proposed quarks in 1964, four years before the first major experiment at SLAC showed fine structure to the proton. They had fairly solid reasons for proposing some sort of substructure to the proton, but no solid reasons for supposing this substructure must be made up of discrete particles. Interestingly, there was not then any competing theory, that I know of: a theory that would propose internal spins or motions instead of particles (as mine does). The understanding of the mechanics of spin has always been so partial that no one ever once suggested any serious counter-theory to quarks, and no one has to this day. The most serious alternative to quarks was Feynman's theory of partons, but it did not differ greatly from the Gell-Mann/Zweig model, and so was soon absorbed by it. Earlier theoretical attempts, like those of Sakata, are also not of the wave variety, relying instead—like all later attempts—on a non-mechanical interpretation of the data.

Because the theory of QCD predated the data, the data was read to fit the theory. This is the way of modern physics, and I have already shown how it has effected the theory of black holes, among other things. But this method must be topsy-turvy. When you have a hammer in hand, everything starts to look like a nail. When you have a new theory—built from near-zero data—everything begins to look like confirmation of it. From the beginning, QCD got very little critique and very much effort at proof. As soon as the major accelerators went online with their new energies, they were put to use seeking confirmation of QCD. If a new particle was theorized, it was immediately sought. If a new particle was discovered, the theory was immediately tweaked to include it. No one wanted to disprove the theory and everyone wanted to prove it. Is it any wonder that it has stood so well? Forty years later, no one is seriously bothered by the non-appearance of the quark (or of the Higgs boson, etc.). Most have accepted that the quark is infinitely unstable, and they will not be bothered if we never find it. And one assumes they will find some way to feel good about the non-appearance of the Higgs boson, even after we exhaust 200 or 300GeV limits. QCD will then borrow the apologetics of string theory, claiming that a particle with the mass of a grain of sand is too unstable to be found. Of course this will not stop them from using tax dollars to build even larger colliders, to seek other hypothetical particles.

One begins to think that, given the lack of critical thinking among physicists and science readers, it is no longer necessary to hide gaping theoretical holes. No one has any desire to see problems, and so they are guaranteed not to. They embrace the non-mechanical whimsies of QCD and dance about, fully prepared to waltz with any pretty paradox. Hence the open-armed welcome of string theory.

QCD is like string theory in another way: string theory has conveniently pushed all possible proof to the 10^{19} GeV energy range, and QCD has followed suit by continuing to push its experimental proofs to higher and higher energies. This trick forces the math to carry all the weight, and that is just how new physics wants it. These energies are permanently out-of-reach, and this is no accident. It gives physicists the excuse to build an infinite line of bigger machines, based only on promises, and prevents any final tally of the theory.

Yes, modern physics depends on its math and its machines. Both deflect current criticism. Regarding the math, QCD, like other modern theories, has doubled and tripled its protection. It has warned physicists away from closely analyzing anything, taught them to do physics without serious logical analysis, and then built a Byzantine series of walls to prevent break-outs. QCD comes complete with its own very high walls. These walls provide insurance in those rare instances that anyone sobers up, pokes his head above ground and opens his eyes. These walls are mathematical walls, of course. As with Relativity and QED, all the heuristics of QCD are couched in a very difficult engineer's math. The math is glamorous in itself, and this deflects much of the potential disappointment or self-criticism from "physicists", who like to be seen doing important things. Since theorists are not really doing theory anymore, and physicists are not really doing physics, by the old definitions, it is good that they can substitute pretty math for pretty theory, achieving equal or greater levels of prestige and causing equal or greater levels of awe among the laity.

As the perfect example of this, see Edward Witten, the top dog now in string theory. Witten's most impressive skill is as a mathematician, and we are told he is the first physicist to win major mathematical medals. This is no accident. The mathematics departments love Witten, since he has continued to buy and import all their newest toys. Physics is not physics anymore, it is math. Like math, it is no longer proved by experiment or simple physical interaction, it is proved by the perceived "beauty of the edifice." Since mathematicians like bigger edifices for their own sake, that is what we now get in physics. With QCD and string theory, physics is becoming "pure." It has long since lost its physical foundation (it lost that with the Copenhagen interpretation) and now it is little more than a flight of fancy, with (at times) a few rules of internal consistency.

I say engineer's math because the math of QED and QCD is full of short cuts. Short cuts in math are good for efficiency, but are terrible for theory. Therefore, highly compressed math is great for engineers, but terrible for physicists. If you are doing mechanics (dynamics or kinematics), you want a very simple transparent math, with variables clearly assigned, no groupings of variables to confuse things, and all postulates right out in the open. But this is not what we see with quantum physics. QED and QCD are both buried under the Hamiltonian, to start with, and we must assume this is no accident. At its simplest, the Hamiltonian ($H = \Sigma pq - L$) is a compression of the Lagrangian with a summation of the generalized momenta and position coordinates. The Lagrangian [$L = \Sigma (T - V)$] is a compression of the kinetic and potential energies. So all individual mechanical variables—time, length, etc.—are buried very deep. In my overhaul, I will not only be throwing out all unnecessary particles and variables, I will be throwing out all unnecessary operators and algebras. The engineers can reintroduce the

Hamiltonian and so on after we finish doing theory, to simplify their calculations. But for now, the math of QCD must be seen as one more wall to tear down.

This also applies to SU(3), which stands for special unitary group, a form of matrix math using Lie algebra. Just as it was never shown why the tensor calculus was necessary to Relativity, and it was never shown why complex numbers were necessary to E/M, and it was never shown why non-Euclidean math was necessary to gravity, it has never been shown why Lie algebra and matrix math is necessary to nuclear theory. In fact, SU(3) is not only unnecessary, it has proved impossible to apply. For example, one of these mathematical holes within the greater hole is called the Faddeev-Popov ghost. This ghost is a type of renormalization, and it requires adding ghosts fields to already calculated gauge fields, in order to make them consistent field theories. What this means is that the fancy math is known not to work. Just as in QED, the equations have to be pushed after the fact in various ways. Consistency is achieved only by fudging, since the data refuses to match postulates in a normal way. But rather than jettison these maths and start over, modern physicists prefer to keep them and plaster them up with chalk and band-aids. After all, it is the math that is the main draw, and if they ditched the pretty math and big book of terminology, 9/10's of the sex-appeal of the field would evaporate. Physicists and science readers aren't disturbed by things like Faddeev-Popov ghosts: they are thrilled by them. A house full of ghosts is so much more exciting than a house full of logic and reason. But these fancy maths are not necessary at all: they were invented in moments of ennui by mathematicians who needed something to do in order to advance their careers. The maths were brought into physics for the same reason. They act as novelty, misdirection, and brainwashing, all at the same time.

If your eyes are on the math, your eyes are off the theory. If you are thrilling at ghosts, you are not analyzing the baseboards and the flooring. In graduate programs, they can't answer your fundamental questions, and they have almost no real theory to teach you, so they must fill their days and your days somehow. They do this by teaching you these wordy and intricate maths. As Feynman bragged, it takes years to learn all the math. Yes, and during those years, you will have become a computing machine, incapable of independent thought or simple critique. With your formidable list of words and manipulations, you will feel secure, so secure that you will never notice that six foot hole next to your bed.

The truth is, like all the other physical problems of history, the problems of QCD can be solved with logical postulates and fairly simple math. Difficult math is not useful at the theoretical level. No, it is a menace, nothing less. I have proved this in my other papers, where I have solved difficult and embedded problems without any modern math, and done it in a fraction of the time. I will do the same here.

I have already shown in another chapter that the strong force is unnecessary. This proves how weak the entire structure of QCD must be. In this chapter I will go beyond this, to show that we can also jettison the gluon, the boson, the weak force, and, yes, the quark itself. In this way, it may be that almost nothing of current theory will remain after our housecleaning but a few bricks clinging together in piles of rubble.

What allows for this overhaul are two fundamental and far-reaching discoveries. These discoveries are revolutionary, but shockingly simple. The first is that the E/M field is always repulsive at the foundational level. I have jettisoned attraction from all physics, and required all theory to be built on physical postulates. That is to say, I have resuscitated the original definitions of dynamics and kinematics. Interactions must be based on motions and forces by contact. It was thought that this could no longer be done, but I have shown that it *can* be done. I have shown that the motions of electrons relative to protons and fields can be explained without attractions, and I will show the same with particles within nuclei.

The second discovery arose from my study of superposition. In earlier chapters I have shown that all the paradoxes of superposition can be explained simply by stacked spins, with each outer spin beyond the gyroscopic influence of inner spins. These stacked spins explained both the wave nature of all spinning particles as well as the famous anomalies of detectors. What I soon realized is that these stacked spins could be applied not only to electron or photons, but also to protons and neutrons. Once we apply these stacked spins to hadrons (baryons and mesons), we are able to explain particle breakups and particle compositions without quarks.

This is because the number of the degrees of freedom we achieve with such an analysis matches the degrees we achieve with quark theory [see just below], and the symmetry also matches current quark theory. Stacked spins replace the current quark quantum numbers, as well as explaining mechanically many things that have so far gone unexplained. Chirality and symmetry are given a physical basis, among many other things.

So these are the fundamental changes we have so far:

1. The mediating photons of charge are not virtual or "messenger." There is no attraction. All charge is repulsive. Charge is mediated by *B*-photons, by straight bombardment. Not all particles must have charge, since not all particles must

emit *B*-photons. Among particles with charge, this charge is a function of surface area. Therefore if we give the proton a charge of 1, the electron no longer has a charge of -1. It has a charge of about 1/1836 or .000545. The *B*-photon is also not mass-less and is not point-like. It has a calculable mass and radius, both of which are about G (6.67×10^{-11}) times the mass and radius of the proton. That is, the *B*-photon is eight million times smaller than the electron.

2. Gravity is a measurable force at the quantum level. It has been hidden in the mis-defined and mis-applied charge field. The total E/M or charge field we now measure at the quantum level is not a single field, but a unified field. Once we un-unify it, or separate it, we find the gravitational field as a major player once more.

3. We have four possible spins on every quantum. A quantum may have all these spins, or only some of them. A quantum that loses an outer spin will seem to change from one quantum to another.

Here is a list of the 16 possible spin states of a baryon.

+a+x+y+z
-a+x+y+z
+a-x+y+z
-a-x+y+z
+a+x-y+z
+a-x-y+z
-a+x-y+z
-a-x-y+z
+a+x+y-z
-a+x+y-z
+a-x+y-z

-a-x+y-z
+a+x-y-z
+a-x-y-z
-a+x-y-z
-a-x-y-z

Notice that we have 16 independent states. That this doubles the so-called "eightfold way"** of chromodynamics is no accident. In the matrix equations of Gell-Mann we have 17 non-zero entries in the matrices (the last matrix had three entries). This is also not a coincidence.

As a start, you may think of the three x, y, z levels as replacing the up, down and strange quarks. The a-level then becomes the color variable. Because x, y, and z must have different energies, quarks appear to have different masses. But quark mass is a mirage. There are no quarks, and the mass differences are energy differences only.

According to this list, mesons are baryons stripped (during collision) of the outer spin. Mesons over the baryon mass are particles that have unstable spins on top of the stable z-spin.

The SLAC experiments in 1968 that showed substructure to the proton were showing not quarks, but spins. As we know, every particle is a wave, and we can now see that it is a wave with four possible inner structures. In a field or during inelastic scattering, inner spins will act like constituent particles, since they will have not only their own directional forces, they will have their own diameters which are less than the particle's apparent diameter.

QCD has translated the three outer spins as three quarks. The axial spin is buried deep beneath the other three spins, and its radius is the actual radius of the particle, so it is "hidden." It is this axial spin that has become "color" in QCD.

The proton and electron create the electrical field, via the linear energy of emission. The z-spin of the emission creates the magnetic field. The y and x-spins of all quanta are relatively buried, and are noticeable only in nuclear interactions. No known macro-forces are caused by them.

At very high energies, you can force other spins on top of the stable z-spin. These spins are highly unstable: the natural mass and radius of the particle cannot support it, and outer spins interfere with inner spins. These outer spins explain the appearance of quanta above the baryon energy. This explains W and Z particles. They are not bosons and are not responsible for any weak force: they are simply baryons with superadded spins. All of the energy for this new spin comes from the field, and in this sense the new particle is a temporary freak, not a mediating field particle.

To show this, let us take a fresh look at beta decay. According to the standard model, a neutron decays into a proton, an electron, and an electron anti-neutrino. It does this by first emitting a W boson. In accelerators, we "see" the first four particles in this decay, but we do not see the W boson. The W boson only lives for about 10^{-27} seconds, we are told, so its track is too short to see, even did it exist (which is convenient, I would say). We believe in the W boson for theoretical reasons, and because the Feynman diagram tells us that it is the W boson that then decays into the electron and electron anti-neutrino. The theoretical reason we believe in it that we want very much to prove the existence of Higgs mechanisms and spontaneously broken symmetries, and we have been pursuing that for forty years (the new hadron collider is more money and time thrown in this direction). The reason we want to prove the Higgs mechanism and spontaneously broken symmetries goes far back into the history of QED, and is basically the attempt the cover our tracks for the past 80 years. In short, since the Copenhagen Interpretation forbade any mechanical talk about quantum particles, we could not provide for the mass of these particles in any classical way. This set up a chain of mistakes and mis-definitions and mis-assignments, which ultimately led to the illogical Higgs field. The Higgs field is a field of virtual particles that underlie or make up the vacuum. For the vacuum to act like the vacuum, these virtual particles must obey symmetry: they can have charge over a very small interval, but the total charge must sum to zero. In the presence of matter, however, the vacuum breaks this symmetry, creating charge that does not sum to zero. This charge creates a force, which creates mass in the matter.

Now, any honest person could see that the Higgs field is non-mechanical and non-physical. It is something from nothing. Besides, it is a clear *reductio*. If you are going to assert that mass comes from nothing, in a bogus, non-mechanical fudge, why not just do it within matter? Why take the extra step and perform your magic trick on the vacuum? Why not perform it directly on matter? Why wave your magic wand over the vacuum, to create a force? Why not wave your magic wand over mass to give it mass, as Newton did? Higgs thought he was doing something clever by backing up a step, but I don't see what he achieved. Honestly, all he achieved was another half-century of busy-work.

But mainstream physicists have been impressed with the Higgs field and gauge theory from the beginning because it allowed them to create a new fudge of their own. This spontaneously broken symmetry was very suggestive to them, since they had a symmetry they needed broken as well, and they couldn't see how to break it. As I have shown elsewhere, this "symmetry breaking" is really just fancy words for "fudging an answer." These physicists had created a rule, convinced everyone of that rule, browbeaten thousands of graduate students into accepting that rule, and now they wanted to break it. In short, they needed to break the flavor symmetry, but chromodynamics didn't allow it. Chromodynamics was symmetrical. To break this symmetry required they create a whole new force (the weak force) and two or three new particles. Once tied to the sexy Higgs field and gauge theory, this fourth fundamental force would be radical enough to go over the head of the old school, and to break the symmetry in an imposing and impressive way.

It looked like working at first, since it won the physicists Nobel Prizes, gave them book contracts, and made at least one of them, Weinberg, a household name. Unfortunately, the weak force and the W and Z particles depended on the Higgs boson, and the Higgs boson still has not been found, 40 years later. With each passing decade, this gets (or should get) more and more embarrassing, which is why we heard so much about the Superconducting Supercollider in the 80's, and why we have to hear so much about the Hadron Collider now. The Higgs boson has now become "the god particle," since according to the standard model, it creates everything—which will be all the more ironic when it turns out to be a ghost. We will need another Nietzsche to tell us that our god is dead. I volunteer, and begin now.

We don't really have to study the electroweak theory or gauge theory or the Higgs field to see how tenuous and illogical this all was from the beginning. We only have to look at the Feynman diagram I described above. We are supposed to believe that the W boson decays into the electron and electron anti-neutrino. And we are supposed to believe that the W boson was emitted by the neutron. The problem is that the W boson has a mass of about 80 GeV/c^2, which is more massive than an iron atom. How is energy conserved in this emission and decay? Where does all that extra mass come from, and where does it go when we get down to the electron and electron anti-neutrino? 80GeV decays down to .5 MeV? That is like saying that a marble emitted the Moon (while remaining a marble), and then the Moon decayed into a grain of sand and a whisper. Feynman's diagram is incomplete, to say the least.

Is there a simpler way of explaining beta decay? Of course there is. We just have to return to my list of stacked spins and decide which one is the proton and which one is the neutron, and then go from there. We begin by taking the two most important facts known to us about neutrons. One, they have no charge; two, outside the nucleus they decay in about 15 minutes. Concerning the proton, we know that it has a charge and that it is stable. I have shown that charge is emission of *B*-photons, so the proton is emitting and the neutron is not. Why is the neutron not emitting? That is our first problem to solve mechanically. We will begin by assuming that all baryons emit equally (from the surface of the particle itself): this will save us from having to make up new rules for no reason. But we see that this emission must travel out from the surface of the particle through four stacked spins. It is clear that some combinations of spins allow the emission through and some don't. Some combinations must block charge emission, and therefore charge. But how could they do this, in simple mechanical terms?

To see this, we simply follow the maze. To simplify the analysis, let us look at only one plane in the problem. We let the particle spin on its axis, and we define the x-dimension as the plane created by extending the equator. In this plane the particle is emitting photons like a pinwheel or spinning water sprinkler.

To simplify even further, let us follow only one dt of emission. This emission will be a circle, obviously, one that is increasing in radius and decreasing in density. Now we add two motions: linear motion of the entire particle and end-over-end spin. To complete our analysis, we only have to add one more thing: the spin of the emitted photons. B-photons have wavelength and therefore spin just like any other particle. In this simplified analysis, we will look only at its outermost or z-spin, which we will define as clockwise.

Can we imagine any blocking at this point? No. In order to block anything with end-over-end spin, the particle would have to outrun its own emission. Since we assume the emission speed is c, this is an impossibility. Can we imagine any difference in +x or -x, that is, in clockwise end-over-end versus counter-clockwise end-over-end? No. There is no way that chirality could cause a mechanical difference at this point. Now we continue to the next level of spin. We encounter a larger end-over-end spin, of magnitude 4, in the y-plane. Can we imagine any blocking here? Again, no. So let us continue. We proceed to the next spin level, of magnitude 8, orthogonal to the others and outside their gyroscopic influence. Can we imagine any blocking here? Yes!

To show this in a simple illustration, let us transpose the fields into one plane, so that it can be drawn in photoshop. Let us try to draw the very first of the 16 possible combinations, that is +a+x+y+z. The initial emission from the surface of the particle will be straight out from the particle, in a line. Then, when we meet the next spin level, we must turn right or left. Let us assign right to "+". In this combination, we have three right turns, which must bring us back to where we were, as you can see clearly.

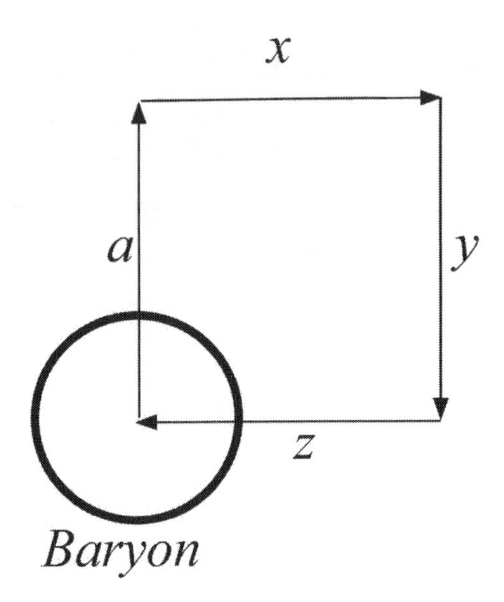

The emission is trapped by the stacked spins. It cannot escape the particle, and so the particle must be neutral. In this one-plane analysis, -a+x+y+z is also a neutron, since the last three turns are again right turns. Three left turns must also forbid emission in the same way, so +a-x-y-z and -a-x-y-z would also be neutrons.

Now, this analysis is a slight oversimplification, since we are only analyzing in one plane; but it shows you the method. If we do the full analysis, we find instead that these states are neutrons:[2]

-a-x-y+z -
+a+x+y-z -
-a+x+y-z -
+a-x-y+z -

They are neutrons for two reasons: 1) the emission cannot escape, and 2) the emission returns with the *B*-photons upside-down. That is what the last minus sign means here. Upside-down photons have a reversed spin, therefore they return opposite in every way to the way they were emitted from the surface of the particle: the energy is canceled completely.

These next states are anti-neutrons, since the returning spin of the radiation is not reversed. That is, the *B*-photons return upside-up:

+a-x+y+z +

+a+x-y-z +

-a-x+y+z +

-a+x-y-z +

Protons are:

+a+x+y+z

-a-x-y-z

-a+x+y+z

+a-x-y-z

Anti-protons are:

+a-x+y-z

-a+x-y+z

+a+x-y+z

-a-x+y-z

With the protons, the emission escapes from the stacked spins in the same state it was emitted from the surface of the particle. It has the same direction and chirality. With the anti-protons, the emission escapes in the same direction as it was emitted, keeping its full energy; but the emission is upside-down. That is to say, it is electrically positive and magnetically negative.

This means that the difference between particles and anti-particles is not a difference in the particles themselves, or in the size of the emission. The difference is in the combination of stacked spins. Certain combinations give us an emission field of *B*-photons that are spinning clockwise. Other combinations give us *B*-photons that are spinning counter-clockwise. When these two fields meet, they must cancel each other's spins, and therefore each other's repulsive energies. The particles emitting will therefore be unprotected: their fields are gone. They are prone to collision.

The neutron should also be prone to collision, since its *B*-photon field cannot get beyond its z-spin. But it is not as vulnerable to collision as some particles, since it retains its spins regardless. These spins are some protection, since they must be destroyed before the particle is "naked". The z-spin has a lot of energy, whether *B*-photons are passing through it or not. This is the mechanical reason it survives for a short time outside the nucleus.

This is also the reason the neutron has more mass: it doesn't lose the energy of emission. Since energy is mass, we may deduce that the mass equivalence of the emission of a baryon is 2.3×10^{-30} kg. The neutron traps this emission; the proton emits it.

Finally, this also explains the slight mass difference between the neutron and anti-neutron, in a direct mechanical way. The emission is trapped by both particles, making them neutral; but in the anti-neutron, the trapped emission does not cancel its own energy precisely. A clockwise spin will cancel a counter-clockwise spin, when two particles of equal mass meet head-on. But if two particles of equal mass meet head-on, and each has a clockwise spin, the spin is not canceled. No, it is doubled. Which means that *B*-photons trapped in anti-neutrons cannot cancel out completely. Only their kinetic energy, or energy from forward motion, cancels. But the spin energy of the emission remains. Depending on which of the four anti-neutrons we are talking about, this spin energy can either augment or tamp down the spin energy of the particle. So the anti-neutron can weigh slightly more or less than the neutron.

This mass difference also ties into the color problem. In his Nobel Lecture, David Gross tells us,

> Color had been introduced by O. W. Greenberg (1964), Y. Nambu (1965, 1968) and M. Y. Han and Nambu (1965). Nambu's motivation for color was twofold; first to offer an explanation of why only (what we would now call) color singlet hadrons exist by postulating a strong force (but with no specification as to what kind of force) coupled to color, which was responsible for the fact that color-neutral states were lighter than colored states.[4]

My explanation above provides a simple mechanical cause of this mass difference, without the need of the theories that have been pasted together since the '60's. To understand both mass and chirality differences, we look at the stacked spins, and the way these four spins channel the emission. Once we understand this very powerful analysis and learn to use it, we don't need the idea of color at all.

Besides explaining mass differences, the stacked spins also explain the magnetic moment of the neutron. QCD cannot explain this in a straightforward manner. As usual, it requires a lot of *ad hoc* theories and new non-mechanical terms and interactions. But my spins explain it simply and immediately. Just study the stacks above, for neutron and proton. A neutron is always a proton with a reversed outer spin. Since the outer spins are reversed, the action in the magnetic field must be reversed. The magnetic field is caused by the spins on the charge photons that make up the field, and these spins must interact directly with the outer spins of baryons. This is why, if we define the magnetic moment of the proton as positive, the magnetic moment of the neutron is negative. But why is the value of the magnetic moment of the proton about 1.5x that of the neutron? Shouldn't the neutron's value be zero, since it is not emitting the charge field? No, the magnetic moment is not a measure of the baryon's own field; it is a measure of how much the baryon reacts to the charge field that is already there. Since the neutron has a radius and a spin, it has a presence in the field. It cannot dodge the charge photons in the field, so we would not expect its magnetic moment to be zero. Its value is lower than the proton because it is not emitting itself, but it is not zero. From the difference in values between the proton and neutron, we may calculate that the proton (and anti-proton) supplies about $1/3^{rd}$ of the charge field at its own surface. This makes $2/3^{rd}$ of the charge field ambient or residual. This will be very important in later papers.

But we still must explain the neutron's actual decay. Why does it decay into a proton and an electron, with a bit of energy left over? Well, study the composition of the neutron and proton: all you have to do is switch the z-spin to make one into the other. So the neutron does not decay at all. No, *it gets hit.* It doesn't emit the electron, it gets hit by a positron in a glancing blow, reversing the z-spin of the neutron. The collision also reverses the z-spin of the positron, turning it into an electron.

Beta decay has been misread. The detectors have failed to detect the incoming particle, and so they think it has been created. This is because the outgoing electron follows the same path as the incoming positron, simply reversing direction. As the spin is reversed, so is the linear direction. This would "overwrite" the incoming track, making it invisible in any detection chamber.

From all this, one could predict that beta decay can be increased by a source of positrons of the right sort. . . except that it is already known. In reverse, this is what was happening with the 1956 experiment of Clyde Cowan, *et. al.* (who won the Nobel Prize for it in 1995). Beta decay was used to promote anti-beta decay. Cowan *et. al.* interpreted this experiment as a proof of neutrinos, but it was only a proof that the proper kind of electron in the proper B-photon field could promote anti-beta decay.

But we still must explain the neutrino. Sadly, there is no neutrino in beta decay. I know that I am taking all your prized particles from you, but that is how it goes. You can hardly complain of the large number of elementary particles, and then also complain when they are shown to be ghosts. Do you want a more elegant theory or do you not? It is like Pauli complaining that if he had known what a mess quantum physics would be, he would have gone into botany; and then adding to that mess. Pauli is responsible for this neutrino, in fact, and we have it because Pauli, like the rest, proposed a new particle to fill every hole. I am showing that we don't need particles to fill every hole. We need logical motions and interactions.

The neutrino was proposed by Pauli to fill an energy difference. The electron plus the proton do not equal the neutron. In fact, we are about an electron and a half short, just as a matter of mass. The neutron is 2.5 electrons heavier than the proton.

But all this is beside the point, since once again the historical analysis is wrong. We don't have too little energy or mass after the decay, we have too much. If we include the energy of the *B*-photon field emitted by the proton, then the proton and the neutron have exactly the same mass. Before the hit, we had the neutron, and the field was internal. After the hit, we have the proton, and the field is now external, being emitted as a charge field. The electron would appear to be over the conservation line, in this sense. We have an entire electron worth of energy too much. But I have already explained that: we simply failed to detect the incoming positron. We have four types of positrons, and we have never detected the positron that causes beta decay (or at least detected its track in a beta decay experiment).

That said, we do have a failure of conservation, even given the incoming positron and the energy in the *B*-photon field. But this energy failure is not 1.5 electron masses. It is much smaller than that. And I will now show you both the size and the reason for the size. The standard model can tell you the first, but not the second. The standard model has no mechanical explanation for any of this.

The energy failure in beta decay is explained very simply by the energy difference between the incoming positron and the outgoing electron. Both have the same mass and the same size charge, you will say. True, but charge is no longer simply plus or minus, as I showed above. Even the charge has spin. What I mean is that the same analysis I applied to the neutron and proton must be applied to the electron and positron. The electron and positron are also emitting a *B*-photon field. It is about 1,836 times less strong than the field emitted by the proton, but it exists and it is repulsive. Yes, both the electron and the positron are repulsing all other particles. In that sense they both have *positive* charges. Look above: both the proton and anti-proton have positive charges, according to my analysis. Both are emitting and therefore repulsing, by straight bombardment. Therefore their opposite natures are not really a function of charge. They are a function of spin: the spin of the *B*-photons that are being emitted. Well, the same thing applies to the electron and positron. Both are emitting and therefore both are repulsing, but the *B*-photons of the positron are upside-down, relative to the *B*-photons of the electron. As I said above, the result of this is a magnetic field difference, not an electrical or charge field difference.

Let that sink in fully, and then go back to the beta decay. The neutron becomes a proton, by switching its z-spin. This switch allows the *B*-photon emission of the baryon to get through. So, before the positron arrived, we had no emission from the big particle. The positron hits it, and it begins emitting. But the total *B*-photon field—existing after this interaction—is made up of both particles, both the baryon and the lepton. The positron's *B*-photons were upside-down, relative to the proton, but the electron's *B*-photons are upside-up. The positron would have subtracted from the field energy, but the electron adds to it. And so the missing energy is not in an invisible neutrino, it is in the total *B*-photon field. The charge of the field has appeared to go up, because the charge itself has spin. The *B*-photons are spinning, and we have to monitor not only the number of photons but also the spin of the photons. In other words, the switch from positron to electron boosts the total angular momentum of the photon field. The amount of that boost is what we call the neutrino. The neutrino is not a particle, it is a field surge. The neutrino is like a step-up in the magnetic moment of the *B*-photon.

This explains the experiment of 1956, since an augmented *B*-photon field will act exactly like a theoretical field of neutrinos.

It will travel and it will carry energy. This augmented B-photon field will have exactly the right energy and spin to facilitate anti-beta decay.

It also explains neutrino oscillation, one of the major thorns of the standard model. Current theory can explain neutrino oscillation (if at all) only with *ad hoc* mathematical fixes. I can explain it with simple mechanics. In beta decay, the field emitted by the lepton inverts, due to the collision, and this changes the energy of the total field (baryon + lepton). Well, a similar thing happens in "neutrino oscillation." Neutrinos don't really change flavor, since neutrinos don't exist. What happens instead is a change in energy of the total B-photon field, and this change is due to collisions. In short, you have a series of interactions. In the first interaction, neutrons are freed from the nucleus in some manner, setting up beta decay and thereby the small rise in the B-photon field that I described above. This gives the experiment the appearance of an electron neutrino field. In the next step, the neutron is stripped of its outer spin in a second collision, turning it into a meson of one sort or another. In this collision, the baryon is stripped of its z-spin completely, and the electron is stripped of its axial and x-spins. This changes the direction of the emission field, but this is not what we are measuring after the second collision. The muon neutrino is not a field fluctuation, but the electron stripped of its spins. It takes three electron hits to strip the z-spin of each neutron, and these three stripped electrons huddle to create the muon neutrino. The muon neutrino has exactly three times the energy of a non-spinning electron, as I showed in my chapter on mesons. So there is no oscillation. There are two separate interactions, neither of which really involves a neutrino.

I realize I have explained only the first and last major problems of QCD, and that there are hundreds more to solve before I convince anyone that my method is superior. But, as I said going in, my intention with this chapter was not to solve every problem or to create a "final theory" of anything. My intention was to critique the current model and to suggest very strongly that a mechanical model could be and should be found. I think I have done that. I have shown that a mechanical model is not only possible, it is much simpler, much more transparent, and much more satisfying as a logical endeavor. Some will find my method naïve, but I embrace this "naiveté." It is what used to be called clarity. It is what used to be called physics.

*The worst, though, are "truth" and "beauty", given sometimes to top and bottom quarks—Keats would be appalled.
**The Buddha would also be appalled to be included in all the dishonesty and misdirection of QCD.
[1] http://en.wikipedia.org/wiki/Color_confinement
[2] To perform this manipulation will require different things for different people. Some will need a computer model to follow the turns; others will need math; others can do it in their heads. I almost hate to admit it, but I used a toy dog. I start with the toy facing front and traveling forward, and I imagine him spinning CW or CCW at each turn, depending on whether I am in a situation with +a or -a. Clockwise I assign to +a, and I do this simply to keep the right-hand rule. According to the right-hand rule, if you are moving forward, you are spinning CW. The right hand rule applies to electricity and magnetism, but I assume that what works for the z-spin must also work for inner spins.
[3] REVIEWS OF MODERN PHYSICS, VOLUME 77, JULY 2005, p. 839
[4] David Gross, REVIEWS OF MODERN PHYSICS, VOLUME 77, JULY 2005, p. 840.
or http://www.qedcorp.com/APS/DGNobelRevModPhys_77_000837.pdf

[5] See my newer paper The Infinite Weakness of the Theory of Weak Interaction for a fuller explanation of how and why gauge math fails: http://milesmathis.com/weak2.html

[6]See my paper on Stern-Gerlach, http://milesmathis.com/stern.html

ON LAPLACE and the 3-BODY PROBLEM

Among other things, Laplace is famous for extending Newton's equations of motion into specific problems in the solar system, like the perturbation between Jupiter and Saturn. It had been known for some time that the orbit of Saturn was getting a bit larger and that of Jupiter a bit smaller. What was required was for mathematicians to match the equations to the data. Euler and Lagrange had made some headway on the problem, but it was far from solved when Laplace tackled it 1776. His first assays were not fruitful, the very first being an attempt to consider the action of an ether. Remember that, because I will show that he was right as a matter of mechanics the first time. There is no ether, but the perturbations are not simply gravitational, either. The perturbations are also electromagnetic.* They are caused by the unified field and therefore cannot be explained by straight pulling forces alone. At any rate, he soon gave up on mechanics and returned to a strictly heuristic and mathematical solution. That is, he did not consider the causes of the forces, he considered only their sizes. He wanted to match the math straight to the data, in the modern way. That is why modern physicists love him.

Euler had at first considered only the mean elongation—the straight-line distance between the planets and Sun. If gravity as a straight pulling force were the only mechanism in celestial mechanics this should have gotten the right answer, which is precisely why these great mathematicians were stumped for so long and why Laplace was looking at an ether 30 years after Euler's failure. Newton's equations don't use or support an ether, but the equations and theory of Newton weren't matching the data, so Laplace was a bit desperate. Euler next tried factoring in the eccentricities, but using Newton's theory, these should not make any difference. The shape of the orbit should not matter in Newton's equations, since it is the distance between the objects that determines all the forces. And, indeed, the eccentricities did nothing to help solve the problem. So Lagrange and Laplace began to look at secular equations, or, in other words, at the magnitude of the remaining inequalities. Again, these should not make any difference if Newton's theory is correct. In fact, using Newton's equations should not provide us with *any* inequalities, since inequalities are defined as irregularities in the orbit beyond the straight gravitational pulls. But Newton does not allow for this. The only way to mechanically account for these inequalities, using straight pulling forces, is to assign them to still more bodies, like Uranus or Neptune, but this is not what Laplace did. Why? Because Neptune was not discovered yet, and Uranus was only just discovered (1781) and they didn't have any good data on it.

Besides, the "great inequality" between Jupiter and Saturn is still not thought to be caused by Uranus or Neptune or any other gravitational perturbations. It is now said to be caused almost entirely by the 5:2 resonance of the two orbits.[1] But as I have already shown in my gravity papers, these resonances cannot be caused by gravity alone. The modern argument is circular: a resonance is just a number relation, a mathematical outcome, and cannot be a cause of anything. The modern argument claims that the resonance is caused by gravity and that the great inequality is caused by gravity, and then that the great inequality is caused by the resonance. That is not a logical line of cause. That is like saying that Jan gave birth to Bob and Jan gave birth to Jim, and that Jim is the cause of Bob.

In current theory, the great inequalities are actually not anything mechanical. The "inequality" is just a name that Laplace made up to give to the gap he needed to fill between what Newton's math could really explain, using the mean elongations, and what the visual data was telling him. It was not an inequality, it was a margin of error. And he filled it just as I claimed in the other paper, by finessing the math. Laplace was a master of this, just like all the other big names would be from then on, from Gauss to Hilbert to Feynman to Witten. A good mathematician should always be able to push a set of equations a few percentage points, and that is all he needed.

Laplace's solution is a clear fudge, because the forces between the objects, if caused by Newton's gravity, can only be a function of the distances between them. The eccentricities, inclinations, and mean motions cannot have anything to do with it, much less the third and fourth powers of these numbers. Laplace, like so many others, uses the calculus only as a trick to allow him these higher terms, by expanding things that aren't really expandable. What I mean is, Laplace doesn't need to look at the equations for the orbits themselves. He doesn't need to concern himself with curves, neither circles nor ellipses. He only needs to know the distances between his three objects, which are straight-line distances. Then he can integrate these over time. This is why Euler's first solution should have gotten the right answer. But by looking at curves, Laplace is able to expand the equations into infinite series, using tricks based on Newton's own binomial expansion (think power series). Any curve equation can be expanded into an infinite series like this, because a curve is based on powers above 1. The curve is a series of differentials and a differential is a binomial.

This trick is still being used today, as I have shown recently with *gamma*. Because Einstein's *gamma* contains a square root, it can be expanded into an infinite series. This infinite series can then be used to show higher orders of numbers, and these orders can be interpreted in various sloppy ways. But since I have shown that *gamma* is derived incorrectly, and have shown that the real transforms do not contain a square root, all the interpretations, now called parameterized post-Newtonian formalisms, are false.

It is the same with Laplace. Laplace interprets these higher orders as inequalities, but they are not inequalities, they are terms in the expansion. You cannot treat terms in the series separately, as if they have some life of their own. Specifically, you cannot assign them to unknown perturbations or to your margins of error. Einstein makes a similar mistake when he assigns the first term in his expansion (which is 1) to Newton's field, and the other terms to relativistic corrections. He can't do that, because in a series expansion the terms may not be read that way. The series is not just as list of terms separated by commas, it is a list of terms that must be added. Notice the plus signs in the expansion!

This is of paramount and fundamental importance, because although Laplace may be able to match data by manipulating these terms over longer periods, in doing so he is obscuring the mechanical cause of the perturbations. Not only is he *not* able to show the mechanics underneath his math, his math acts as a heavy blanket, keeping future physicists and mathematicians from questioning what his math tells us about the field it is representing. And it tells us some very strange things indeed!

Before I show those strange things, let me clarify the previous paragraphs. Laplace's first trick is to completely ditch Newton's own equations and theory. Laplace doesn't use Newton's equations, or Kepler's either. Why should he, since they are known to fail? No, he makes up his own equations, based upon the observations themselves. In Forest Ray Moulton's book from 1914, which is published in full on the web by Google Books[2], you can see (p.202) that Laplace uses the first and second derivatives of the direction cosines, and these angles are taken directly from the three or four real observations.

$$d^2x/d\tau^2 = -x/\tau^3$$
$$d^2y/d\tau^2 = -y/\tau^3$$
$$d^2z/d\tau^2 = -z/\tau^3$$
$$x = \rho\lambda - X$$
$$y = \rho\mu - Y$$
$$z = \rho\nu - Z$$

[I refer to Moulton's book because, besides being completely orthodox on this question, it is easy to find, whereas the English translation of Laplace by Bowditch is out of print, unavailable on the internet, and otherwise difficult to find.] In other words, the angles aren't based on calculations from Newton's equations or theory, they are taken directly from the data. If this data doesn't fit Newton's theory (and it doesn't), then Laplace will have directly bypassed the theory at this point. From the very first step, Laplace has detached his math from Newton, attaching it instead to the data.

You will say, fine, good for him. As scientists we are interested in the data, first and foremost. Yes, Dr. Feynman, but if you have bypassed the theory on your first step, you cannot later claim to have confirmed the theory, can you? If the real observations were in strict agreement with Newton, Newton would not have had to propose a hand of God, Halley would not have shown a problem in the orbits, and Euler would have had no trouble finding a solution. Laplace was working on the problem precisely because the data did not match the prediction of gravity as a simple straight-line pull in the three-body problem. Laplace's solution is a solution of the data, not of Newton's theory. Laplace finds a way to solve the equations, by discovering the finer resonances they contain, but because his initial equations were taken from the data, he cannot, at the end, claim to have discovered a solution that completely confirms Newton or that proves the stability of the solar system. In fact, he has done precisely the opposite. He has proved that the three-body problem can be solved only by going well beyond Newton's math and postulates and theory.

My critic will say, "What do you mean he attached the equations to the data? Didn't he just create equations to *contain* the data, like anyone else would?" No, look at his equations. All you have is accelerations and directional cosines. You don't have any representation of the fields present, and Newton's theory is a field theory just as much as Einstein's. Newton's gravity is a mass field, since his masses determine all forces and the forces determine the accelerations. Just imagine that you wanted to fit in my charge field here, so that the final accelerations were caused by not one but two separate fields acting in tandem. You couldn't do it, because Laplace has already begun differentiating from the first step. The acceleration is a single differential in each dimension, so that it can be differentiated or integrated in only one manner. The second derivative of x, for instance, cannot be manipulated in two different ways at the same time. But if each acceleration is caused by two forces, and these forces vary in different ways, then Laplace will have closed out any possibility of correct mathematical representation of his force fields. Even in the case that he brings mass back into the equations later, he still will not be able to represent all the variations in the fields. This may be why he gave up on the idea of the ether early on. You see, he doesn't need an external ether, acting independently of his mass field. He needs his "ether" to already be inside his equations from the first step, because Newton's ether in already inside his equations (unknowingly). As I have shown, Newton's main gravity equation works in most situations because it already contains the E/M field (and the E/M field is a sort of ether, roughly). This is why he needs G in the equation: it acts as the scaling transform between his two fields. But Laplace can represent none of this, since his equations are not proper field equations. He doesn't have enough complexity in his initial representations of the orbits to show all the field variations. In modern terms, he doesn't have enough degrees of freedom. The real field in the solar system has two acceleration fields, not one, even before you get to relativity, and Laplace has no way to represent that. His equations aren't even as good as Newton's, since Newton has not one but two masses in his main equation. Because he starts with these masses, *not* written as derivatives or partial derivatives, we can split each mass into volume and density variables. This immediately gives him twice as many degrees of freedom as Laplace has here. We can give volume to one field and density to the other, so that (in most cases) both will be represented in the final solution. This is precisely why

Newton's equation works in real life, despite the fact that no one has ever been able to unwind it.

I will now show the oddest thing in the entire history of this problem. It is fantastically odd, not only in itself, but because no one in history, not even Laplace or Lagrange or Euler, ever made comment on it, to my knowledge. If they ever saw this fact underneath their numbers, they must have swept it under the rug. Or perhaps the fact has been excised from history, by one group or another. My mother (who has a PhD in mathematics, and who scored 800 on her math GRE in 1960, when the 99th percentile for women was 650) taught me to approach all math problems by first estimating a ballpark solution. Whenever possible, I should look at the problem intuitively or spatially or in a commonsense fashion, and this would keep me from moving toward answers that were clearly illogical. In other words, look at the mathematical problem without math, to start with. Let us do that with this problem of Jupiter and Saturn. Let us strip it of all math and all complications, and see what we would expect, using the pretty simple postulates of Newton. Let us look at four positions of the two planets, in four diagrams, in a straight 3-body problem.

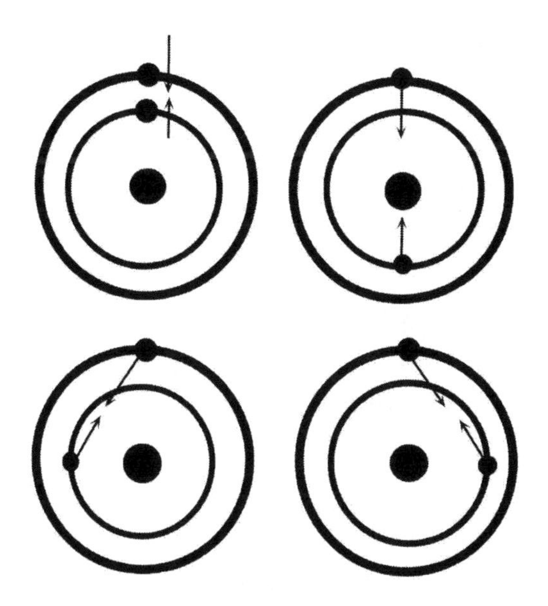

With gravity as a force of attraction, and no other force fields playing a part, we find that we would expect the gap between Jupiter and Saturn to be getting *smaller*. Because, while Jupiter is sometimes being pulled away from the Sun by Saturn and sometimes toward it, Saturn is *always* being pulled toward the Sun by Jupiter. I challenge you to find a relative position of the two planets where Saturn is not being pulled into a lower orbit by Jupiter.

In fact, Saturn could not be pulled into a higher orbit, even over a single differential, unless Jupiter went above it. In other words, the orbit of Jupiter would have to be so highly eccentric that Jupiter's semi-major axis would have to exceed that of Saturn's for some part of their combined orbits. And even then, Saturn's orbit would be increasing over only a small span of years. It could not increase, as a whole, from the time of Chaldea or Ptolemy until the time of Laplace. It is clear from the simplest analysis that Saturn cannot go higher due to the laws of Newton, without the influence of outer planets. But Uranus and Neptune were not included in Laplace's perturbation math. It is strictly impossible for Laplace to have proved what he is said to have proved.

For this simple reason, you can dismiss any and all claims to the contrary, no matter how much impressive math or argument they throw at you. You cannot prove something that is impossible, and if someone does prove it, you can be sure they cheated. You can spend a lot of time finding the specific cheats, or you can just use the simple proof. They cannot have proved it, because you have already proved that it is impossible.

My critic will claim that it is quite easy to imagine the gap between Saturn and Jupiter getting larger or the orbit of Saturn increasing, even given my figures. Since the orbit of Saturn is slightly elliptical, we just need to go to Saturn's nearest approach to the Sun. The differentials after that will all show a larger orbital distance, no matter what Jupiter is doing. Yes, but that is not what Halley found, is it? Halley was not talking about a short-term increase in the orbit of Saturn, during one of its 30 year cycles. Halley was talking about an increase over centuries. My simple diagrams show that the only long-term change in Saturn's orbit we could expect, as a function of a pulling perturbation from Jupiter, is a going lower. The only perturbation from Jupiter that would send Saturn higher over the long term is repulsion. That is so clear from the first look that I can't believe I am on this page having to say it. In no part of any long-term cycle could the orbit of Saturn, as a whole, appear to increase, given gravity alone.

All Laplace did is assume long-term stability of the solar system, and then find a math to match the given data. But this is a double cheat, 1) because gravity by itself cannot explain the given curves and motions, and 2) because this assumption of long-term stability is clearly false. Just as Laplace could not have proved that Newton's equations explain the perturbations of Jupiter and Saturn, he could not have proved stability, since there is no stability. To this day, this question of stability commands much time and effort and verbiage[3], but all you have to do is look at the asteroid belt to answer it. If the solar system is stable, why do we have so many obvious indications of collision, and this magnificently obvious proof of major collision?

The standard model in this area now claims that the asteroid belt was never anything but collection of debris left over from the solar disk or nebula. Their main argument in this regard is that the debris we have studied varies in chemical composition, so it could not have come from a single planet. The combined mass of the debris is also low, so it would have to be a very small planet. But neither of these arguments is strong at all. If we pulverized the Earth or Moon down into tiny fragments, we would also find chemical variation, since the core differs greatly from the crust. The asteroid belt is likely the remains of at least two planets, since we propose a collision, not a natural breakup. Planets vary greatly in chemical composition. As for the low combined mass, that is also easy to answer, since a large part of the mass must have been ejected immediately from the area in collision. More would have been lost over time to collision and ejection, so that even if the collision were fairly recent, we would expect only a fraction of the original masses to remain in the area. But the best argument for a breakup of a planet already at that orbit is the low amount of collision we see today. The residual pieces are amazingly stable at that orbit and trajectory, and that is probably because they were already at that orbit and trajectory to start with. This is the residue that avoided great momentum changes during the first impact. Planetoids and debris herded into this area by Jupiter would be prone to collision, unless they were all herded from the same area at the same time. And debris left over from the nebula would have to be incredibly stable to have come down to us in this fashion, staying in a primordial state while all around them evolved into open space or major planetary orbits or satellites. Finally, the model now states that the planetesimals in the asteroid belt were too perturbed by Jupiter to form a planet; but Jupiter's own Moons, perturbed both by Jupiter and by Saturn, are not too perturbed to be large and round and orderly (large and round compared to the planetesimals, anyway). Rather than herding the planetesimals and preventing them from combining, it is much more likely that Jupiter has stolen many of them, making them into its smaller satellites. No, modern theory on the asteroid belt is just another example of the de-evolution of science and rationality. Until the last fifty years, the shattered planet hypothesis was widespread, since it is so obvious, and only now, when we cannot stand the idea of instability, are we forced to retreat into counterintuitive models.

The great mystery of the asteroid belt is Ceres, which is nearly perfectly round and amazingly pristine. It is like nothing else in the belt. Most likely it was the distant Moon of the planet hit, and thereby avoided the brunt of the first collision. To explain its non-ejection from the main orbit upon the loss of its primary, we only need to imagine it was moving parallel to the primary at impact, in one of two orbital positions. And its composition may give it a high electromagnetism, which would repel other slow-moving intruders in the belt. It is already proposed to have a high water content (as ice) and water is a very good conductor, as is known.

But back to Laplace. Can we propose a solution? Well, we know from observation that Saturn is in fact going higher, at least over part of its long-term journey. And we know that gravity by itself cannot account for this. Which means that something else must be causing the resonance and the long-term motion. We must have gravity PLUS something else. This is why Laplace first looked at the ether. He must have known what I have just proved, and he wanted to solve it without leaving Newton completely in the bin. An ether would have been calculable as a straight addition to Newton, and Laplace could then hope to match the pair to the data, truly saving Newton. Besides my many papers on the foundational E/M field, which clearly show the E/M field already existing within the equation of Newton, physics has long had ample data that the planets have strong E/M fields, easily strong enough to be included in perturbation theory. Magnetospheres, solar wind exclusion (with and without a magnetosphere), plasma research, tangential torques on planets and satellites by their primaries (unexplainable by gravity, which cannot work at a tangent), and so on. In the 50's it was discovered that Jupiter was sending out strong radio noises (predicted, amazingly enough, by Velikovsky), and more recently similar noises were found coming from Saturn. You can hear this "Saturn singing" by going to ABC, which has a copy of NASA's own tape.[4] Beyond that, it has long been known that the Moon strongly affects radio reception, causing a complete doubling. Also in the 50's, John H. Nelson of RCA showed that the planets strongly affect radio transmission as well, and that a clear "electrical balance mechanism" was at work among all the bodies of the solar system. All these signals, noises, and effects are direct evidence of electrical perturbations. If the mainstream have ignored them or buried them, that is not proof they do not exist.r a long space of years.

It is claimed that these electrical effects aren't strong enough to enter Laplace's or Newton's equations, but I have shown that they are. Using simple calculations and postulates, I have shown that the charge perturbation from the Moon to the Earth is about .151 m/s^2, or about .46% of the total or unified force between them. That is far from negligible. Moreover, the charge perturbation is positive or repulsive. Yes, the foundational E/M perturbation is *always* opposite to the gravitational force, and this must affect all the terms in any equations of motion.

My critic will say, "Even if you are right in all this, we still can't send Saturn higher unless the E/M component of the unified field is shown to be larger than the gravity component. To send Saturn higher requires a straight repulsion, and as long as the gravity component is larger than the E/M component, the unified field must remain attractive, as a total force. You have shown that the E/M component is much smaller than the gravity component. So your theory does not help us. The unified field must still always be attractive." That is logical as far as it goes, but it doesn't take into account the nature of the E/M field. Yes, when we look at highly stable and nearly round orbits like that of the Moon, the E/M field is not doing a lot of the work. The Moon is moving quite fast at a good distance, and the effective E/M repulsion between the Earth and Moon can be quite low. But I have shown that the E/M field is capable of very quick response and enlargement, since it varies to the fourth power, not the square. In other words, if we bump the Moon into a lower orbit all of the sudden, say half of its distance now, the E/M field repulsion between the two bodies will immediately increase by a factor of 16. This difference in field variation between the E/M field and gravity is precisely what gives the orbit its "float." As I showed in my chapter on celestial mechanics, real orbits have much more stability than can be accounted for in Newton's or Einstein's theories. Even Kepler's theory of ellipses cannot work without a large degree of float or correction. Neither circular orbits nor ellipses can be explained with pulling forces alone, since a balance of two motions or forces (gravity and tangential velocity) cannot be stable. It requires three motions to create stability over a series of intervals. Because the push (E/M) varies more quickly than the pull (gravity), we have three motions that can correct themselves against each other, creating long-term stability.

This solves the problem of Saturn, because we can now explain the resonance in a logical and mechanical fashion, without using fancy math to hide it or big names to shoo off analysis. It was known even in the time of Halley that this divergence in the orbits of Saturn and Jupiter meant that they must have been quite near each other millions of years ago. If we reverse time, the two orbits must get nearer and nearer. Laplace showed that they wouldn't have collided, because the resonance causes the orbits to begin separating again, after a long space of years. He showed that what we have is a dance, where they are moving away for half of the long period and moving closer for the other half. All good, except that using his math, and the theory of gravity, you can't explain *why* this is. As Jupiter and Saturn get nearer, they should attract each other more, not less, right? It is the mechanics that causes the math, not the math that causes the mechanics. The math is a representation of physical causes, the math is not the cause itself. So we must have some reason that Jupiter and Saturn decide to start moving away from each other again, despite being brought together. Newton, Kepler, Laplace, and Einstein can't explain it, which is why it has been buried up to this day. No one wants to admit that the problem exists, so they pile century after century of math on top of it, to hide it. Even Laplace's equations weren't thought to be sufficient, so Poincaré is used to create a deeper pit.

But the solution is quite simple, once you know that the E/M field is available, and that it changes to the fourth power.

When you bring Jupiter and Saturn near to each other, the E/M repulsion kicks in very strongly, and the closer they get, the stronger it gets. It bounces them apart, in the same way the Sun bounces a comet away. Gravity cannot create such a "well" on its own. So-called gravitational wells are real, but they can only be caused by the unified field, or the E/M field and gravity working in tandem, as I show in my chapter on the ellipse. This explains the point of greatest divergence of the orbits of Jupiter and Saturn as well, because as the orbital gap increases, the repulsion of the E/M field shrinks rapidly. A quick shrinking of the E/M field strength is the same as a quick building of the unified field strength, and Jupiter is able to pull Saturn back in line, assuring that Saturn's orbit does not continue to grow indefinitely.

[1] http://www.daviddarling.info/encyclopedia/I/inequality.html
[2]http://books.google.com/books?id=0ZwRAAAAYAAJ&pg
=PA429&dq=laplace+celestial+mechanics&lr=#v
=onepage&q=laplace%20celestial%20mechanics&f=false
[3]http://www.scholarpedia.org/article/Stability
_of_the_solar_system
[4]http://www.abc.net.au/science/news/stories/s1425596.htm

BODE'S LAW

Bode's Law, also known as the Titius-Bode Law, is one of the most famous unexplained laws in the Solar System. The first mention of the law is from 1715, so this one has been sitting around without a solution for almost 300 years. Since that time, the form of the "law" has been:

$a = n + 4$
where $n = 0, 3, 6, 12, 24, 48....$

Wikipedia glosses for us the standard model explanation:

> There is no solid theoretical explanation of the Titius–Bode law, but it is probably a combination of orbital resonance and shortage of degrees of freedom.

The first explanation is just a guess, but it is a bad guess since orbital resonances have been given to gravity, but no one has ever shown a mechanical cause of any "gravitational resonance." Resonances cannot be caused by gravity, and no one in history has shown that they can. I will show that the failure of Bode's law on outer planets is caused by a charge "resonance", but this resonance has little to do with the main series of numbers and nothing to do with gravity. The second explanation from Wiki is an even worse guess, since it turns the truth on its head. What this editor means by "shortage of degrees of freedom" is that the math falls into this alignment because it can't do otherwise. But the failure of the standard model to explain Bode's Law is actually caused, in part, by a shortage of degrees of freedom in another sense. Current celestial mechanics lacks an entire field, since it hasn't incorporated the foundational E/M field into its equations. I have shown that the foundational E/M field or charge field is already in Kepler's and Newton's equations, and this field may be called a mathematical degree of freedom. The equations of celestial mechanics lack the required complexity, so they can only be pushed after the fact, as with Laplace.

Even worse than these two wild guesses is what comes next at Wiki.

However, Astrophysicist Alan Boss states that it is just a coincidence, and the planetary science journal Icarus no longer accepts papers attempting to provide 'improved' versions of the law.

Well, may Alan Boss take a flying leap into a pit of cankered pions and may *Icarus* catch a malign meteor in the teeth. If this isn't a case of science being shut down by fiat, I don't know what would be. These people and institutions should be shunned by all real scientists and thinkers, and I recommend that *Icarus* be put into bankruptcy, by readers refusing to read it. I have never read it and never will. If these editors will be intellectually bankrupt, then they should be financially bankrupt as well.

Unfortunately, science is now controlled by this sort of person, and it isn't just *Icarus* that is the problem. All of mainstream physics is now like this. We have "scientists" that have stated and published opinions prejudicially on things they know nothing about, and they don't seem to understand that this is unscientific. Alan Boss can't explain this phenomenon and he wants to be sure no one explains it while he is working on something else. Infantile. He probably throws a fit when his wife works on the crossword puzzle when he is out of the room. Physics has been taken over by very small people.

Since I am the one solving these longstanding problems, it is not for me to follow their policies. The ignorant do not set rules for the wise. If any of these journals wants to publish real papers instead of fake papers, they had better change their attitudes. They seem to believe it is my loss, not being able to publish with them, but it is their loss. If Einstein had not published with *Annalen der Physik*, would it have been his loss or theirs? *Annalen* gets a historical mention only because of Einstein.

Remember that, *Icarus*. Signed, Apollo.

In physics, as in all else, problem solvers are primary and publishers are secondary. Publishers are just administrators and committees and gatekeepers. Historically and scientifically, they are of no import. Publishers can refuse to publish, but they cannot stop the spread of information.

Wiki also tells us this:

Dubrulle and Graner have shown that power-law distance rules can be a consequence of collapsing-cloud models of planetary systems possessing two symmetries: rotational invariance (the cloud and its contents are axially symmetric) and scale invariance (the cloud and its contents look the same on all length scales), the latter being a feature of many phenomena considered to play a role in planetary formation, such as turbulence.

There it is again, the power law used to fudge a very squishy answer. I have already shown in the chapter on Laplace how the power law is used to push bad equations to fit data, by using infinite series to assign terms to errors. These errors are called "remaining inequalities" or something like that, and a lot of fancy math is used to drop or add terms. But since I will show that Bode's law can be explained without any sort of calculus, with 9th grade algebra, all this talk of collapsing clouds is just further nebulosity. These are the types of papers the mainstream likes to publish: papers full of pompous impossibilities like scale invariance (clouds are density fields and cannot be scale invariant) and incomprehensible and undefined equations. The mainstream physics paper is little more than an institutional efflux, a career propellant adding another anti-logical pollutant to the infinite stream of modern absurdities.

As you will now see, the solution to this problem is so simple that it makes three centuries of physicists and mathematicians look like bumblers. I looked at that sequence of numbers for about half a minute before I saw it was based on the square root of 2. The "law" has been in the wrong form since the beginning, and so no one was able to see the proper sequence.

Currently, the sequence goes like this:

4, 7, 10, 16, 28, 52....

But it should be written as

$4, 5\sqrt{2}, 7\sqrt{2}, 11\sqrt{2}, 20\sqrt{2}, 36\sqrt{2}$....

Which can be written as

2^2

$(2^2 + 1)\sqrt{2}$

$(2^2 + 1 + 2)\sqrt{2}$

$(2^2 + 1 + 2 + 2^2)\sqrt{2}$

$(2^2 + 1 + 2 + 2^2 + 3^2 + 4^2)\sqrt{2}$

If we want to express this with Mercury as 1, then we just divide by 4.

$2^2 / 2^2$

$[(2^2 + 1)\sqrt{2}]/2^2$

$[(2^2 + 1 + 2)\sqrt{2}]/2^2$

$[(2^2 + 1 + 2 + 2^2)\sqrt{2}]/2^2$

$[(2^2 + 1 + 2 + 2^2 + 3^2)\ \sqrt{2}]/2^2$

Which expands to:

$2^2 / 2^2$

$\sqrt{2} + (1/2^2)\sqrt{2}$

$\sqrt{2} + (1/2^2)\sqrt{2} + (2/2^2)\sqrt{2}$

$\sqrt{2} + (1/2^2)\sqrt{2} + (2/2^2)\sqrt{2} + (2^2/2^2)\sqrt{2}$

$\sqrt{2} + (1/2^2)\sqrt{2} + (2/2^2)\sqrt{2} + (2^2/2^2)\sqrt{2} + (3^2/2^2)\sqrt{2}$

Which simplifies to:

1

$(5/4)\sqrt{2}$

$(7/4)\sqrt{2}$

$(11/4)\sqrt{2}$

$(20/4)\sqrt{2}$

You will say, "Great, you expressed Bode's Law in terms of $\sqrt{2}$. So what?" Well, the so-what is that it ties directly into my correction to Newton's orbital equation $a = v^2/r$. I have shown that the equation should read $a = v^2/2r$, since our current expression of the orbital velocity is not a velocity. Yes, $a = v^2/r$ works if $v = 2\pi r/t$, but $2\pi r/t$ isn't a velocity. It is a curve over a time, which isn't a velocity. It is just a heuristic ratio that we like because it is easy to measure. But since the orbit curves, it must be an acceleration, and that acceleration is expressed by the equation,

$$a_{orb} = 2\sqrt{2}\pi r/t$$

If we use that acceleration along with the corrected equation, $a = v^2/2r$, we get the same relationship we have now with $a = v^2/r$ and $v = 2\pi r/t$. That is where the $\sqrt{2}$ comes from. And that is why our current orbital equations work despite being faulty. They are consistently faulty, so we don't see the faults except when we are in need of mechanics like I am doing now. Our current heuristics doesn't prevent us from doing engineering, but it does prevent us from doing foundational mechanics like this, or from solving simple problems like explaining Bode's law.

All this has been ignored historically, so we have heuristic equations that have been hiding a lot of important information. Our orbital mechanics, like our celestial mechanics, has been a compressed engineer's math, instead of a theorist's math.

You will say, "I still don't get it. How does the 2 help us solve this?" It helps because, via my corrected orbital equation, it tells us that Bode's law comes right out of the orbital equation. Look at my expression of the series above. We have the square law from Newton's orbital equation, but it is an additive square law. Each planet is not only orbiting the Sun, it is also orbiting inner planets. The Earth's relative distance is given by this equation:

$$r = [(2^2 + 1^2 + 2^2)\sqrt{2}]/2^2$$

In that equation, the Earth is the third term in parentheses, Mercury is the first, and Venus is the second. So we are being told that the Earth's orbit is added on top of the other two. The Earth is orbiting them as well as the Sun. You have to include the inner orbits in the equations for the outer orbits. Very simple, right? In this way, the corrected Bode equation is an analogy of the Pauli exclusion principle. But unlike with Pauli, here we have the mechanics in full view right in front of us. The Earth cannot share Venus' orbit, because the Earth's term must be added to the term of Venus. The Earth is excluded from the second orbit, because each term is added, as a distance, in a set pattern. "Two squared" cannot be at the same distance as "one squared", since the numbers determine the distance.

And it is not just the math that provides the exclusion. As I have said elsewhere, math is just an expression of the mechanics. Math cannot be a cause. The mechanical cause of this math is the exclusion provided in the unified field by the charge component of that field. Gravity cannot exclude. Only the charge field can exclude. The Earth cannot inhabit a different part of Venus' orbit because the two bodies have different charge fields. The Earth has a greater charge field than Venus, due to greater mass and density, so the Sun keeps it at a greater distance. The Sun's charge field and the Earth's charge field actually meet, in space, physically, particle to particle.

If that is all I had to say, this paper wouldn't have been terribly convincing to the hardliners, I admit. I have shown the law as a function of the square root of two, but unless you were impressed and convinced by my correction to Newton's equation (and that is unlikely), this will not seem very momentous. The reason this spins out into a good story of its own is that expressing Bode's law with the square root of two allows us to correct it. We have been told that Bode's law fails with Neptune and Pluto, but predicts the other orbits pretty well. I will now tell you exactly why Bode's law fails with Neptune and Pluto.

Bode's law fails with Neptune and Pluto because Bode's law is in the wrong form. It mimics the right progression by a sort of accident. This is the only possible way that Alan Boss can be seen as correct. Bode's law, as written, is nearly correct only by accident. It is a mathematical coincidence that it follows the right progression for the inner planets. Titius and Bode and the rest just matched a simple math to the data, without any mechanics, and their math is not complex enough to fit the real mechanics. It wasn't even transparent enough for us to see that it was coming straight out of the orbital equation $a = v^2/r$. I have solved that problem, but I still have to show that, although my corrected "law" is superior to Bode's, my failures can be corrected. Yes, the greatest difference between my law and Bode's is that I can show you why the planets outside Jupiter fail to fit the pattern. I will give you the pattern, show you the variance, then show you the correction to the variance, solving the problem completely.

In the current Bode list, Jupiter's number is predicted to be 52 and Saturn's is 102. That would be expressed by me as $36\sqrt{2}$ and $72\sqrt{2}$. Unfortunately, Saturn doesn't fit my pattern. In my pattern, Saturn should be $(36 + 5^2)\sqrt{2} = 61\sqrt{2}$. Then Uranus would be $(61 + 6^2) = 97\sqrt{2}$ and Neptune would be $(97 + 72\sqrt{2}) = 146\sqrt{2}$. Or, if we take Mercury as 1:

1
$(5/4)\sqrt{2}$
$(7/4)\sqrt{2}$
$(11/4)\sqrt{2}$
$(20/4)\sqrt{2}$
$(36/4)\sqrt{2}$
$(61/4)\sqrt{2}$
$(97/4)\sqrt{2}$
$(146/4)\sqrt{2}$
$(210/4)\sqrt{2}$

As a first approximation, that would be the series I would predict with my math. But my series diverges from Bode's series at Saturn. It fails at Saturn rather than Neptune. Why?

Before I tell you, let me show you that my series already matches the data as well or better than Bode's series. I predict an orbit for Venus of 1.024×10^8. The actual orbit is 1.075×10^8. An error of 4.77%. Bode's law predicts an orbit for Venus of 1.034×10^8, an error of 3.8%. I predict an orbit for the Earth of 1.433, an error of 4.2%. Bode's law predicts an orbit for the Earth of 1.478, an error of 1.2%. I predict an orbit for Mars of 22.52, an error of 1.18%. Bode's law predicts an orbit for Mars of 23.64, an error of 3.73%. I predict an orbit for Ceres of 4.095, an error of 1.04%. Bode's law predicts 41.37, an error of .02%. I predict an orbit for Jupiter of 73.7, an error of 5.32%. Bode's law predicts 76.83, an error of 1.3%.

My average error for the first six planets is 2.75%. Bode's average error is 2.1%.

For Saturn I predict 124.9, an error of 12.85%. Bode predicts 147.8, an error of 3.11%. For Uranus, I predict 198.6, an error of 30.97%. Bode predicts 289.6, an error of .66%. For Neptune, I predict 298.9, an error of 33.6%. Bode predicts 573.3, an error of 27.3%. For Pluto, I predict 430, an error of 27.2%. Bode predicts 1,141, an error of 93%.

So, for Saturn and Uranus and Neptune, Bode is better. For Pluto, I am better. My average error for the four outer planets is 26%. Bode's average error is 31%. But my errors are grouped much better, since they go from 13 to 34, a deviation of 10.5 points from my mean. Bode has a deviation of 46 points. I have no complete misses, while Bode's prediction for Pluto can be called a complete miss.

Just as a matter of statistics, my equations for the ten planets are better than Bode's. My average error is 12.1%, while Bode's is 15.2%; and my deviation is much less. Even so, I admit that my correction is not completely convincing at this stage. A 33% error on Neptune is not impressive, for instance, unless I can show the cause of the error. So I will now show how my errors can be taken down to almost nothing, with simple mechanics. The first thing to notice is that Jupiter causes the predictions to fail. With Bode's law, this was not the case. Bode's predictions are just as good from Jupiter to Uranus as they are for the planets inside Jupiter. Bode's prediction for Saturn is very good, and for Uranus it is astonishingly good. But my predictions go from an average error of 2.7% inside Jupiter to an average of 26% outside. This will prove to be a plus for my series, since Jupiter IS the physical cause of this variance. Bode should have found a variance beyond Jupiter, and he didn't. Beyond that, I will show that Bode's match on Uranus was just a fluke.

Actually, you can see the variance beyond Jupiter even without doing any math. A passing glance at the Solar System would tell you that Jupiter is a dividing line. Inside Jupiter, most things are caused by the Sun. Outside Jupiter, most things are caused by Jupiter. The fields outside Jupiter cannot be the same as those inside.

This became ever clearer to me in my recent papers on axial tilt, where I did the charge field calculations for the outer planets. These four planets determine the unified field variations in the entire System, and cause the bulk of the tilts, both inside and outside Jupiter's influence. And, due to its mass, Jupiter's charge influence is crucial.

If we look at the four Jovians and list their charge field densities with Uranus as 1, we get Neptune as 1.523, Saturn as 3.544, and Jupiter as 22.84. The charge fields of all the other planets are negligible compared to those four (although Pluto will play a small part in the solution below). The charge field plays only a small part in the unified field inside Jupiter, which is why my predicted numbers work until we are past Jupiter. The outer planets all perturb each other strongly, as a matter of charge, and so we shouldn't find orbital distances that can be predicted without the charge component of the Unified Field. What is the easiest way to show this, with the simplest math?

We must first look again at my errors for the Jovians in my predictions, above. What would it take to prove my errors were not accidents? My errors for the Jovians are 5.32, 12.85, 30.97, and 33.6. If my theory is correct, then the perturbations among the Jovians should be shown to follow that sequence of numbers. What are the charge forces between the four Jovians?

As I have shown in previous papers, the relative strength of the charge field can be calculated by multiplying the mass of a body times its density. This is because we seek a charge density, and charge is dimensionally the same as mass. This is one of the secrets of physics: the statcoulomb reduces to the same dimensions as mass, and the Coulomb is just mass per second. I say that this gives us a relative density, because by multiplying mass and density, we can get the charge field strength of one body as a percentage of another body. But we cannot find an absolute amount of charge that way.

We will look at Saturn first. Jupiter has a charge field that is 6.445 times greater than that of Saturn. To find the potential difference between them, we let the charge fields meet at Saturn. If Saturn's charge field is 1, Jupiter's is $6.445^{1/4} = 1.593$. The result is .593. We apply that potential difference in both directions, so the variance at Jupiter is 1.593. But since Saturn is smaller, it will feel a greater variance. It will feel 6.445 x .593 = 3.822. Saturn feels 3.822/1.593 = 2.399 times the variance of Jupiter.

Uranus and Neptune will also perturb Saturn, but their effects are much smaller. I will calculate Uranus' variance on Saturn to show this. Saturn has 3.544 more charge density than Uranus, so if Uranus' charge is 1, Saturn's is 1.372. But if Saturn's field is 1 (see above paragraph), then Uranus' is .729. Uranus is 2.2 times further away from Saturn than Jupiter is, so Saturn will feel a variance of only $.729/2.2^4 = .0311$. We can add that to the variance from Jupiter, obtaining 3.822 + .0311 = 3.8531.

Now Neptune. Saturn has 2.328 times more charge than Neptune, so if Neptune's charge is 1, Saturn's is 1.235. But Neptune is 4.69 times as far away from Saturn as Jupiter is, so Saturn will feel a variance of only $.81/4.69^4 = .00167$. We add that to the other variances, 3.8531 + .00167 = 3.855.

Saturn has 14,743 times as much charge as Pluto. So if Pluto's charge is 1, Saturn's is 11.02. But Pluto is 6.84 times further away than Jupiter, so that force drops to $(1/11.02) \times (1/6.84^4) = .0000415$. This brings the total variance to 3.855.

But we have to add the variances on Jupiter from Uranus and Neptune. Uranus has 3.544 times less charge than Saturn and is 3.2 times further away. Which gives us an extra variance of .00269. Neptune has 2.327 times less charge than Saturn and is 5.688 times further away. So an extra variance of .000411. Adding those to 1.593 gives us 1.596. Then, 3.855/1.596 = 2.4154. Saturn has **2.4154** times the variance of Jupiter.

To see if this has filled our margin of error, we consult the earlier numbers. My error for Jupiter was 5.32%. My error for Saturn was 12.85%. That is a ratio of 12.85/5.32 = **2.4154**. If we compare the two bolded numbers, we find a perfect match.

Before we move on to find the variance for the other planets, let us pause to show why Jupiter's variance is 5.32%. I need to show that in order to finish my proof. I need it because showing Saturn's relative variance is not enough. Saturn's number depends on Jupiter's, so I need to prove Jupiter's number in order to set my baseline. The math is very simple and can be shown in only a few lines. We already know that Jupiter's relative charge density is 22.84. We compare that to all the charges from the Jovians as they exist *at the distance of Jupiter*. We can estimate this by looking only at charge from Saturn, since the other charges are almost negligible. $22.84 + 3.544^{1/4} = 24.2$. Jupiter's own charge is 94.4% of that total charge, leaving a difference of 5.6%. That is (roughly) the cause of Jupiter's variance.

As Jupiter is pulled higher, all the Jovians are pulled higher, as a group. Jupiter sets the baseline and the other Jovians follow. This is precisely what my math shows. I was able to dissolve the entire error for Saturn because I found the variance relative to Jupiter.

Now let us look at the variances on Uranus. If Uranus' charge is 1, Saturn's charge at Uranus is $3.544^{1/4} = 1.372$. Uranus will feel 3.544 x .372 = 1.318 from Saturn.

Now Neptune's variance on Uranus. Neptune is larger than Uranus and outside it, so our previous math is difficult to apply. We will solve by an easier math. Neptune's charge is 2.327 times less than Saturn's, so Neptune's variance on Uranus is also 2.327 times less. Saturn's variance on Uranus was 1.318, so Neptune's is 1.318/2.327 = .5664. But Neptune is 1.128 times further away, so the variance drops to $.5664/1.128^4 = .350$. Because Neptune is larger and outside, this variance is negative: -.35.

Now Jupiter's variance on Uranus: Jupiter has 22.84 times as much charge as Uranus, so if Uranus has a charge of 1, Jupiter has a charge of 2.186. Uranus' variance relative to Jupiter would be 22.84 x 1.186 = 27.09. But Uranus is 1.454 times further away from Jupiter than from Saturn, so Uranus only feels a force of $27.09/1.454^4 = 6.061$.

And, finally, Pluto's variance on Uranus. Uranus has 4,160 times more charge than Pluto, so if Pluto's charge is 1, Uranus' is 8.03. But Pluto is 2.1 times farther from Uranus than Saturn is. So, $(1/7.03) \times (1/2.1^4) = .0323$.

Adding the four variances gives us 7.061. But now we have to scale Uranus' variance up to Saturn's. In the equations for Saturn above, we let Saturn equal 1; here we let Uranus equal 1. To do this we simply use the variance between Saturn and Uranus, from above. It was 1.318, so 7.061 x 1.318 = 9.307. Uranus' relative variance is 9.307 and Saturn's was 3.855. Dividing, we get **2.4143**.

So we return to our prediction errors above. The error for Saturn was 12.85% and for Uranus 30.97%. The ratio is **2.4101**. The bolded numbers again show a very good match. Here we have an error of .00173. We have brought our Bode series error down from 30.97% to .173%.

Not only have I solved the problem I set out to solve, I have found another problem to solve later. Notice we have the same number between Saturn and Uranus as we had between Jupiter and Saturn. 2.4143 here and 2.4154 above. That cannot be a coincidence. We will look at those two variances more closely in the future.

Also, because I have shown that the charge field fills the gap between prediction and data, Bode's raw prediction for Uranus must have been a fluke. Bode's prediction was only .66% wrong, which looked impressive until I showed how the Unified Field caused the orbital distance mechanically. Bode's series is based on a straight extension of the equation $a=v^2/r$, and I have just shown that cannot work by itself. It cannot work because it ignores the E/M field entirely. Since Bode's original math contains no mechanics or mechanical postulates, it must have achieved the correct number by luck.

Now, the set of equations for Neptune. We will start with Pluto, as the nearest planet. Neptune has 6,334 times more charge than Pluto, so if Neptune's charge is 1, Pluto's is .000158.

Jupiter has 15 times more charge than Neptune, so if Neptune's charge is 1, Jupiter's is 1.968. So the variance on Neptune is 15 x .968 = 14.52. But Jupiter is 2.647 times further from Neptune than Pluto is, so we find $14.52/2.647^4 = .2956$.

Saturn has 2.327 times more charge than Neptune, so if Neptune's charge is 1, Saturn's is 1.235. The variance on Neptune is 2.327 x .235 = .547. But Saturn is 2.18 times further away from Neptune than Pluto is, so its variance drops to $.547/2.18^4 = .0241$.

Again, rather than compare Neptune to Uranus, we will solve the perturbation between them by comparing Uranus to

Saturn. We just found Saturn's raw variance on Neptune to be .547. Since Uranus has 3.544 times less charge than Saturn, it's variance upon Neptune must be .547/3.544 = .1543. But Uranus is 1.159 times farther away than Pluto, so $.1543/1.159^4 = .0855$. And again, this variance is negative: -.0855.

Add them all up to get .2344. Once again, we have to scale Neptune's variance up to Saturn's. Saturn is 1.887 times farther from Neptune than from Uranus, so we can develop the transform like this: 3.544 x 6.445 x 1.887 x .2344 = 10.1. That is Neptune's scaled variance. We compare that to Uranus' scaled variance, which was 9.307. Dividing gives us **1.0855.**

According to my series errors, my error for Uranus was 30.97%; and for Neptune, 33.6%. Therefore, Neptune should have been perturbed **1.0849** times as much as Uranus. The two bolded numbers are a near match again. The error is .000533. My method has successfully dissolved the series errors for the Jovians, using very simple math. I think we can confirm that the charge field fills the variances extremely well.

Now let us look at the variance on the Earth. Notice that we are not calculating the Earth as a percentage of any other planet, as we were doing with the math of the Jovians. Instead, we are looking for an absolute motion in the field, as we did with Jupiter above. The Earth has 13.05 times the charge of Mars, so if Mars' charge is 1, the Earth's variance from Mars is $\sqrt[4]{13.05} = 1.9006$. The Earth has 1.3 times the charge of Venus, so the variance from Venus is $\sqrt[4]{1.3} = 1.0679$. Venus is smaller and lower, so it will pull the Earth lower, as Uranus does with Neptune. This makes Venus' number negative. Now the variance from Jupiter. Since Jupiter is larger and higher, its variance on the Earth will also be negative. We compare Jupiter's variance on the Earth to Mar's variance on the Earth. Jupiter has 997 times as much charge, but it is 8.14 times as far away. $997/8.143^4 =$ .2268. Since Mars' charge was the baseline 1 in this paragraph, Jupiter's variance is just 1 x .2268. Mercury will also give us a negative variance. Mercury has 1/14.14 times the charge of Venus and is 2.213 times farther away from the Earth. That is .00295. If Mars is 1, Venus is 10.034, so we multiply by ten. The variance from Mercury is then .0296. We will also include Saturn, for good measure. Saturn has 154.7 times the charge of Mars and is 16.64 times farther away. $154.7/16.64^4 = .00202$. Again, negative, and not completely negligible. Add them all up. 1.9006 - 1.0679 - .2268 - .0296 - .00202 = .5743. We have given the Earth a charge of 13.05 in this math, so we compare 13.05 to .5743. The total charge at the Earth of the Earth plus variances is 13.05 + .5743 = 13.624. 13.05/13.624 = .9578. 1 - .9578 = .04215. Or 4.215%. The error from my Bode series was 4.2%, so we are very close.

Now to answer some important questions about the math and mechanics. You will say, "How can Saturn pull Jupiter higher? I thought you said we didn't have any attractions in your unified field?" Good question. As you can see from my math, I am calculating charge differentials. Using these, we find that planets are pushed higher either by larger planets below or smaller planets above. Conversely, planets are pushed lower by larger planets above or smaller planets below. With this simple rule, we see that Jupiter will push Saturn higher, and Saturn will also push Jupiter higher. They both go higher. But why is that, as a matter of mechanics? Haven't I said that charge is a straight bombardment, which would imply that bodies can only repel each other via charge? Well, yes, in the simplest case, that is true. If two bodies aren't already in the field of a third larger body, that is true. They could only repel each other. But that isn't the situation we have here, is it? To look at Jupiter and Saturn, we must be aware that they exist in the greater field of the Sun at all times.

I did not calculate absolute charge above, as I have already admitted. I calculated relative charge, or a charge differential. I calculated the four Jovians relative to each other, leaving the Sun out of it. But if we want to find a motion in the field relative to the Sun (inward or outward), we have to bring the Sun back into the question. In the closed system of Jupiter/Saturn, I found that we have "more charge out." It is clear why that would move Saturn out. But the charge is not just moving at Saturn, it is moving across the whole system. Charge is moving out at Jupiter as well. If we put that closed system into the greater system of the Sun, then charge will be moving out in the entire vicinity. Even charge below Jupiter will be moving out, because our J/S system has created a low pressure across the entire J/S system.

It is very useful to think of charge like wind, creating low pressure and high pressure. This analogy is much more useful than the idea of potential or pluses and minuses, in my opinion, because you can visualize the field blowing from one place to another. This isn't just a metaphor, it is what is physically happening. The charge field is a very fine particulate wind, and it moves from high density to low density. Density and pressure are the same thing, in this regard. So the charge wind is blowing Jupiter just as much as it is blowing Saturn, and it is blowing in the same direction. For this reason, they both go higher.

You will now say, "Since you mention that paper, why is your math different there than here? In your axial tilt papers you let the outer planet increase as the charge goes in. Here you don't. For instance, you keep Saturn at 1 in the equations, and take the fourth root of Jupiter's charge. In the tilt paper you increase Saturn's charge by the distance."

Another good question. All you have to do is notice that in the tilt papers I am calculating perturbations against the Sun. Here I am calculating the perturbations against each other. There, I have three bodies, with one in the middle. Here, I have two bodies, and no middle. So although the math looks similar, it isn't really the same.

Conclusion:

I have completely solved Bode's law, and I think we are due for a name change for that law. Both Titius and Bode failed rather spectacularly to apply the right equation. The solution is rather simple, as you see, and there was no reason for this to sit in mothballs for 300 hundred years. Physicists couldn't look at it without scales on their eyes, since they had bought the "gravity only" interpretation. Laplace "solved" the perturbation equations 230 years ago, and no one has had the gumption to look closely at them since then. Mathematicians failed to solve this, too, and we may assume it is because they got deflected in about 1820, or 190 years ago, by new maths. They weren't interested in simple algebra like I do here: they wanted to use curved fields and infinities and complex numbers and quaternions and lord knows what else. Actually solving a simple problem of mechanics was beneath them. It really makes you wonder how anything ever gets done.

In physics and math, nothing much does get done, as I have shown. The history of physics and math has not been a wonderland of brilliance and fast progression; it has been a shocking wasteland of deflection, misdirection, and complete incompetence, and it is only getting worse. I expect the response to my papers to continue to be vicious, since there is nothing more reactionary than a field of sinecures. It will be like trying to overthrow the Aristotelians or the French Academy or any other nest of nepotism and privilege and corruption. But they had best put on their waders, because the water is high. I am coming right at them, and I am used to deep currents.

THE EASY SOLUTION TO THE NEW SATURN ANOMALY

In November of 2008, Lorenzo Iorio published an announcement with ArXiv[1] that E.V. Pitjeva had analyzed Cassini spacecraft data, finding that the precession of Saturn did not conform to the predictions of GR. The gap between data and prediction was calculated to be .004-.008 arcsec/cy. The announcement in ArXiv corresponded to a paper by Iorio and Ruggiero in SRX Physics[2] at about the same time, in which the authors showed that theories of long range modified gravity (LRMG) could not explain the gap. This was in preparation for a paper by Iorio in July, 2009[3], which proposed that an undiscovered planet X was causing the gap.

This is a drôle repeat of history, since in 1859 Le Verrier proposed the planet Vulcan as the solution to the original precession problem with Mercury. But we need neither a new planet X nor any LRMG. The only "modification" we need to gravity is the simple one I have shown in my long paper on the precession of Mercury. We don't need to "modify" gravity at all:

we only need to correct the simple mistakes in the field equations we have. In fact, my finding that Einstein's field equations are 4% wrong in the field of the Sun is enough by itself to solve the Saturn anomaly. The accepted value for the precession of Saturn is .1836 arcsec/cy.[4] If we multiply that by .04, we obtain .007. That is nearly in the middle of the range calculated by Pitjeva, as you see. I solved the problem before I even knew of it. To put it another way, I *predicted* in 2007 that Einstein's field equations were 4% wrong. In 2008, that prediction was confirmed by data from Saturn.

I will repeat the short math I did in the other paper. Einstein's field equations are field equations of mass, as everyone should know. To solve, Einstein basically does a mass transform in a curved field. In other words, he takes his mass transform from SR and imports it into his tensor equations. The problem with this is that gravity is not a mass, it is a force caused by an acceleration field. Einstein needs to transform a force, but he only transforms a mass. Since by the classical force equation $F = ma$, force is measured in Newtons (kilogram meter per second squared), Einstein needs to transform mass, length, and time squared all in the same equation. I showed that in a given event, Mercury's mass would increase 1.57 times, while its length would increase by 1.04 and its time would decrease by 1.04. Since $F = ma$, our aggregate transform is just $F = (1.57)(1.04)/(1.04)^2 = 1.51$. The difference between 1.57 and 1.51 is 4%. Since Einstein is only transforming mass, his field equations must be 4% wrong across the board.

Critics will now say that my solution is just another LRMG, or a modified gravity, but it should be clear that it is not. One, because it is not an external modification to GR. It is simply a correction of an internal mistake. Two, because it is not an ad hoc addition to the field. I just showed you why my correction works, and none of the other LRMG's can do that. All the other "theories" are just gap fillers, created to answer specific shortcomings. My mathematical correction is a general solution, complete with all the mechanics. Nor does my solution fall to Iorio's SRX Physics paper, since my solution, although general, has to be applied to each known precession separately. My 4% correction applies to each problem, not to sets of problems. To be specific, Iorio and Ruggiero do the math for the perturbations between sets of planets, as in ΨJupSat = 1.36 ± .06. They find different values for different planet pairs, and conclude that LRMG's can't explain this variance. It is true that LRMG's can't explain this, but I can. Although perturbations cause precessions, calculating perturbations and calculating a change in precession on a single planet are two different things. In finding the number .04, you can see that I am calculating a general correction to the field. To apply that general correction to a specific perturbation between given planets requires more math. Specifically, it requires taking into account the two masses and the distance between them. Because the distances and masses are not equal, we should not expect the field errors or corrections to be equal. Iorio and Ruggiero's equations rule out the logarithmic correction and the power-law correction to gravity, but don't rule out my correction. In fact, their equations confirm my correction, since their tables 13 and 21 show that the variance in perturbations is a factor of distance and mass.

[1] http://arxiv.org/abs/0811.0756

[2] http://www.syrexe.com/physics/2008/968393.html

[3] http://arxiv.org/abs/0907.4514

[4] http://farside.ph.utexas.edu/teaching/336k/lectures/node128.html

STRING THEORY
The Inelegant Universe

Between foolish art and foolish science, there may indeed
be all manner of mischievous influence
-John Ruskin

Most readers will assume I have nothing to say about the math of string theory. They will assume that since I am neither an insider nor a famous mathematician, the subtleties of 11-dimensional math are beyond me. And since the first part of this paper attacks the theory and not the math, many will assume that I am just making a philosophical critique. They are quite mistaken. In Part II of this paper I will make a foundational critique of the math, revealing some astonishing facts that even the princes of the theory will not want to miss. So if you tend to nod off at any stretch of sentences that fails to contain a number or a variable, there is something for you here, too. The big laughs are in the first part of this paper, but the lasting interest lies in the last part.

Part I

The Theory

Since the late 1980's string theory has continued to gain in popularity, until now it has become a sort of fashion. Brian Greene puts it this way in his book *The Elegant Universe*: "[In 1984] there was a pervasive feeling among the older graduate students that there was little or no future for particle physics. The standard model was in place and its remarkable success at predicting outcomes indicated that its verification was merely a matter of time and details. . . . [Then] the success of Green and Schwarz finally trickled down even to first-year graduate students, and an electrifying sense of being on the inside of a profound moment in the history of physics displaced the previous ennui."

Most will find nothing particularly revealing in this quote, I imagine. No doubt Greene believes he is just stating a fact, not baring his wicked soul. But I find in it the entire explanation for the movement in science in the 20th century. The keyword is "ennui". In the late 20th century it took a lot to interest the top graduate students like Brian Greene. They could see no quick road to fame by studying the boring past. What was wanted was an avant garde math or theory to latch onto. This is what had made Einstein famous, and after him Feynman and Hawking and all the rest. Mathematics had been the key, and it looked to continue to be the key in the near future. For Brian Greene and the other ambitious young physicists of our time, the job is not to try to discover why the old avant garde maths aren't working; no, the job is to create ever newer avant garde maths that are harder to test. This will automatically provide fewer empirical contradictions and thereby a stronger theory.

In this paper I will use *The Elegant Universe* as my scratching post. I do this for a number of reasons, but the main reasons are: 1) It is a recent bestseller and has done as much as any book to popularize the theory, 2) It describes an almost unbelievably *inelegant* universe, 3) It is as transparent as thinnest glass, setting me up for easy scores on almost every page. As far as the last reason goes, I will show that it is probably a mistake for avant-garde maths and theories to allow themselves to be presented to popular audiences, especially if the presentation is in a clear language. Brian Greene is a good science writer: good in the sense that a reader can penetrate what he is saying. But science used to understand that obscure theories should always remain in obscure language. That was the only hope for them, no matter the audience. An honest presentation of a dishonest theory is too dangerous. For one thing, it allows other scientists like me to find the flaws too easily. Fully cloaked in its armor of equations, it is not so easy to sort out, even for a mathematician. But stated baldly it becomes a sitting duck.

I find it astonishing that string theory has made it this far. Greene says that the early years were a bit of a struggle, but I don't tend to believe it. The fact that a theory that is such a magnificent mess is on its feet at all is a very bad sign. It shows the uncritical nature of our milieu, not only in the public and publishing sector, but at the highest levels. The reason for this is clear: graduate students like Greene were well-trained in being uncritical, and they have been for more than half a century. The old uncritical graduate students are now deans and department chairs, and they are all very far gone down the road of non-discrimination. The list of things they have accepted at face value is long and shocking. Greene's first five chapters are a public airing of all the absurd things he has accepted without much analysis. It is clear that he has accepted them because he never really cared if they were true or made sense or not. He, like the others, has from the beginning judged each incoming piece of information based on its likelihood to add to his prestige, and anything that was already a settled question could not help in that area. What he and the other ambitious theoreticians were looking for all along was the end point. "Get me to the end-point as fast as possible." Because then they could begin making their personal contribution. "Put me as close to the front of the line as you can, where I can begin pushing."

For these brightest students, physics was no longer seen as a field they could add to, it was a field they could trump. Their greatest goal was to make all of the past immediately obsolete. Basic physics was digested like a breakfast at the drive-thru, Relativity was duly cut and pasted, and QED was memorized by rote. All this was done by the age of 24 or 25. Another year of all-nighters provided them with the latest hyper-maths and theories, so that they could immediately begin discussing ten-vector fields with full abandon at the coffeeshop and braintrust.

In this way science has become just like Modern Art. The contemporary artist and the contemporary physicist look at the world in much the same way. The past means nothing. They gravitate to novelty as the ultimate distinction, in and of itself. They do this because novelty is the surest guarantee of recognition. The contemporary artist always has his nose to the wind, sniffing the air for the next trend. As soon as he gets a whiff of it he is off running. He is always in a race with time, for it is no longer a matter of being best, it is a matter of being first. He therefore congregates with others of his type. They mass at the same hotspots, antennae erect.

The contemporary scientist is the same. He is a social creature, always trying to impress. Rigor impresses no one in the modern world, so he does not even have to fake it. What impresses are lots of difficult equations, with lots of new variables and terms. The ultimate distinction is coining new words for the new math and the new objects. Calabi-Yau shapes and 3-branes and orbifolding: that is rich beyond anything.

The art departments have long since dismantled the old schedules: painting and sculpture are passé, studio art a dinosaur, drawing from the nude a sexist embarrassment. The physics and math departments will soon follow suit, no doubt. Mechanics and kinematics will be jettisoned as a theoretical nuisance, a blockage of creativity. Classical algebra and geometry will become an elective, taken only by historians and archivists. Instead, seventh graders will be offered "The Rudiments of Chaos Theory" and "Fun with Tensors" and "Computer Modeling with *i*."

Let me now show you a few examples of the absurdities that the standard model teaches. I do this to prove that by accepting these absurdities, it encourages a proliferation of more such absurdities. It teaches the graduate students, by example, that mathematical fuzziness pays and that conceptual rigor does not. Let us start with the "messenger particle,"[1] a relatively new beast in the physical zoo. The messenger particle is a photon that tells another particle whether it should move away or move near. The messenger particle was invented to solve the problem of attraction. At some point it became clear to physicists that attraction couldn't logically be explained by a trading of particles. Their old blankets over this problem had begin to wear thin, so they needed a new concept. Enter the messenger particle. With the messenger particle, we no longer have to be concerned with explaining physical interactions mechanically. We don't even have to imagine that movement away in a field is caused by bombardment, which was such a simple concept.

No, we can now explain both movement away in a field and movement toward in field as due to information in a messenger particle. This simultaneously explains both positive charge and negative charge. How easy: the photon just tells the particle what to do. Why did we not think of that before?

Once you accept that quantum particles are on speaking terms, physics is so much tidier. There is no end to what we can explain this way. We can have the particles trading recipes, emailing each other, SMSing, watching TV. It is a theoretical goldmine.

Gluons, weak-gauge bosons, and gravitons are also messenger particles of their various forces. The problem of attraction is solved once and for all, for all possible fields. Gravity is not curved space or a physical force. It is a *commandment.*

The next absurdity is one of Feynman's famous absurdities.[2] This one concerns letting an electron going through the two-slit experiment take all possible (infinite in number) paths simultaneously and then summing over these paths to find the wave function. Any idiot can see that this is just a mathematical consideration and has no physical implications, but Feynman was a special kind of idiot. He insisted for some reason that the math was the physics, and all the special idiots since then have taken his word for it. They love to quote or paraphrase him, as Greene does, "You must allow nature to dictate what is and what is not sensible." Which means, "You must allow me (Feynman) to dictate what is and what is not sensible. I am smarter than you are and if you don't allow me to dictate to you, I will browbeat you mercilessly." Even now that Feynman is long in the grave and incapable of personally browbeating anyone, the special idiots still quote and paraphrase and bow to his authority. Feynman himself was bowing to the authority of

Heisenberg and Bohr, who first decided, by fiat, that the math of quantum mechanics was the physics. Or perhaps he was only learning from their example. Counterintuitive fiats had made them famous with all the toadies, why not make a few of his own counterintuitive fiats and toadies?

Greene tells us outright: "Quantum mechanics requires that you hold such pedestrian complaints [about things making sense] in abeyance." What could be more convenient for a scientist? He is now in the position of a priest. The priests have always said the same thing to non-believers. "You must not expect it to make sense. You must have faith. Trust the Lord." Trust Feynman. He is smarter than you and understands what you should believe. He has filled the blackboards with Hamiltonians and has cracked safes. He has earned the right to say ridiculous things, like the Dalai Lama or the Buddha or the President.

This is the most important thing that string theorists have learned from quantum mechanics: you do not have to make sense anymore. Any contradiction can be relabeled a paradox, any infinity can be relabeled an axiom, any absurdity can be given to Nature herself, who is an absurd creature, in love with illogic and caprice.

I could go on indefinitely, listing other absurdities like the Twin Paradox and the singularity and so on, but I have analyzed these problems elsewhere in great detail; and besides, you already either accept them or don't accept them, so that my comments are nearly beside the point. You won't judge the concepts based on anything I could say of them, you will only judge me for what I say of them. Therefore, let me proceed to critique string theory, a theory that is not yet set in stone, even for the toadiest.

String theory begins by defining a string. In most instances a string is a one-dimensional loop, we are told. String theory is famous for its ever-increasing number of required dimensions, so that you would think that the theorists would have a pretty tight idea of what a dimension is. But if you think this you would be wrong. String theory is about math, not about concepts, and these brilliant mathematicians don't have a very clear idea what a dimension is or what a one-dimensional "thing" would be. In math, a one-dimensional thing is a line. It always has been, since the time of Euclid, and that has not changed recently. A zero-dimensional thing is a point, a two dimensional thing is a plane, and a three-dimensional thing is a cube or sphere or whatnot. But all of these things are mathematical abstractions. They don't exist and can't exist. Of all these mathematical things, only the three-dimensional things have a potential existence, and then only if you add time. There is a very simple reason for this that has nothing to do with gods or turning on the universe or anything else esoteric or metaphysical. Points, lines and planes cannot exist because they do not have any physical extension. A plane disappears in the z direction, a line disappears in the x and y direction, and a point disappears in all three directions. In mathematical terms, it means that the variable or field has hit a limit—a zero or infinity—at this point in the equations, making existence impossible.

Physicists used to understand simple concepts like this, but no more. Even mathematicians don't appear to understand them. These concepts just get in the way until some self-described genius somewhere finds a clever way around them, and we aren't bothered with them anymore. After that we are allowed to propose the existence of mathematical objects and no one blinks an eye. But it remains a (perhaps unpleasant) fact that a line cannot exist. Even in pure math, a "one-dimensional loop" cannot exist. A one-dimensional loop is false even as a mathematical abstraction. Why? Because a loop curves. Any curve is no longer one-dimensional. A curve is two-dimensional, by definition.

Greene and his heroes imagine that because you can, in a pinch, express a position on a curve with one variable, that it is a one-dimensional object. But it isn't. Greene proves this when he begins talking about his garden-hose world, where the position of a bug on the hose can be expressed with two variables. He then admits in an endnote that if the garden hose has an interior, we must have more dimensions. But when, in a physical situation, is it possible to imagine a garden hose, no matter how tiny, with no interior? It is not possible and his "two-dimensional" garden hose, if physical, must have three dimensions.

Greene makes the current confusion even more apparent when he begins increasing the Type IIA coupling constant.[3] This allows strings to expand into two and three-dimensional objects. He says that the two-dimensional string is like a bicycle tire and the three-dimensional object is like a donut. So Greene thinks there is a dimensional difference between a bicycle tire and a donut! If a bicycle tire is not solid rubber through and through, then the third dimension has disappeared? We should at least have to suck the space out of it with some kind of space vacuum, right?

String theory is such a godawful mess right from the first concept that it is painful to go on. But I will. Once we have our impossible one-dimensional loops, we are to imagine that they are vibrating. To vibrate in the right way for the theory, they must be strung very, very, very tight. Now, a sensible person would already have several foundational questions. First of all, why are they vibrating? Second, why are some vibrating one way and some vibrating another way? Third, what causes the tension?

The first concept, basic vibration, we can give them. Vibration is far from being a basic motion, but there has to be some first cause, and so we will allow one unexplainable motion as first cause. But the difference in different vibrations cannot be uncaused. We cannot allow it to be a postulate. Different vibrations should have different mechanical causes. If one string is vibrating in a different way from another, there must be a reason. String theorists have already told us that strings are not made up of subparticles; they are absolutely indivisible. They should therefore be undifferentiated. Ultimate strings that are indivisible should act the same in the same circumstances. If they act differently, then the circumstances must differ. But we are not told what these different circumstances are. The vast variation in behavior is just another postulate.

Besides, even if we admitted the impossible—that a one-dimensional loop could exist—once you give it a vibration it automatically gains a dimension. All you have to do is look at the direction it is postulated to vibrate in. Does it vibrate lengthwise? Of course not. How could it? It is undifferentiated lengthwise, meaning that it is not made up of sub-particles. There is no way a pulse could travel lengthwise in a string that was not divisible. So the theorists propose sideways vibrations, of different sizes and wavelengths. In technical terms, we are talking about transverse waves, not longitudinal waves. A transverse wave will automatically push the string into a second dimension. So all talk of one-dimensional strings is a wash from the beginning, for two fundamental reasons, not one.

This brings us to another question: is it even possible for a one-dimensional string to vibrate sideways? I have reminded the reader that a longitudinal wave is impossible to imagine without some subdivision of the string. There has to be some sort of longitudinal variation to propose compression; but this variation is not possible without subdivision. In the end this is because without subdivision you cannot insert any space into the string. You need space in between the particles making up the string in order to propose variation in compression. But a closer analysis shows the same problem with transverse waves on a one-dimensional string. How is a one-dimensional string bendable without some "give" between particles making up the string? If the string is absolutely indivisible and undifferentiated, then it is not clear that we can bend it. A bend would occur at the bond between particles, in a macroscopic string. In a string-theory string, there is no bond between particles, since there are no particles making up the string. Bending or vibrating a string-theory string is like proposing to bend or vibrate a cube or a cone or a sphere. If our fundamental particle were any of these instead of the string, you would laugh if someone proposed that it bent. Imagine a cube bending. How would a fundamental, undifferentiated cube bend? Or a fundamental, undifferentiated sphere? But bending a fundamental, undifferentiated string is just as silly. It is just another postulate that is impossible to explain or justify.

Likewise, tension is a pretty complex concept. It is not a fundamental motion or event. In fact, tension is a force. But string theory is supposed to be explaining the four fundamental forces, not creating more. What causes the tension? How is it possible to have a tension across an undifferentiated ultimate string? How is it possible to have tension in a closed loop, unless that loop is being expanded by some outside force? None of this is explained. Tension is just an assumption, another axiom.

After a first reading, I had already discovered that string theory has more basic postulates than any theory I had ever seen or imagined. To any logical person from past centuries, string theory would look like a comedy of errors. Newton has been all but laughed at by string theorists for not giving a mechanical explanation of force at a distance. But these theorists are in no position to throw stones. Newton would look at string theory and say something like, "Well, of course, if you are allowed to make enough unprovable assumptions at the beginning, you can formulate a theory to contain anything. Especially if you are allowed to beg the question so egregiously. String theory is the attempt to unify the four basic force fields. To do this it creates, as a postulate, a huge force of uncaused tension. Then it adds to that a basic 'particle' that can morph into anything, just by changing its 'tune.' All these morphs are uncaused and act as further postulates—as postulates they do not require proof or any justification. Then, whenever the math stops spitting out numbers they want, they postulate new branes, donuts, tubes, three-holed buttons, frisbees, and anything else that tickles their fancy. None of these new objects has to be justified beyond the fact that they needed them to fill a hole in the math. 'It fit the hole, therefore it must be real!' Then, when the going gets really tough, they add a new dimension. M-theory gives them the 11th dimension, and why stop there? I predict that, like Feynman, they will finally understand that the sky is the limit. Why not predict an infinite number of curled up dimensions, sum over them in some fudgy way, and achieve any answer you like, to fit any occasion. Only then will the madness come to its illogical end."

This is the basic technique of string theory: if you run into some dead-end at any point in the math, transport that dead-end back to the string. For example, perhaps you find the need for a new particle but your math at that level of size or theory does not allow it. Well, simply make it another axiom of string theory. Postulate that your basic string takes that shape under the circumstances you have discovered, and your work is done.

In this way, every conceivable problem can be collected at the foundational level and made into an axiom. Since you don't have to prove axioms, you will never be pestered to supply a proof or explain anything. All problems can be collected, reinserted at the axiomatic level, and treated ever after as assumptions. In this way string theory really is the perfect theory. Using this technique, nothing is beyond mathematical expression.

String theory is actually even more inelegant than QED, and QED is not exactly a poster child for elegance or simplicity. Greene tells us that string theory was invented to simplify the huge number of "elementary" particles in QED, as well as to combine QED and Relativity. But he seems oblivious to the fact that string theory has a record-setting number of axioms and an ever-increasing number of vibrations, dimensions, blobs, branes, and jellies. The only object not yet incorporated into string theory is the moss-covered three-handled family gradunza. It also has a truly impressive number of manufactured manipulations, such as the set of instructions for orbi-folding a Calabi-Yau shape or the tearing of space in a flop transition. These manipulations come provided with no theory, and are basically added to the list of postulates: postulate #89,041—we can flop-tear an orbi-folded 3-brane goofus as long as we can say afterwards that the math made us do it (and provide a sexy little computer-generated diagram).

Another of the inelegances of string theory is the required energy of a string. The unbelievable amount of tension [10^{39} tons] on a single string gives it a mass of some 10^{19} protons. This is about the mass of a grain of dust. The theorists need all this force on the string since they have gathered all the other forces here at the axiomatic level. This has the added benefit, they think, of making the mass too great to be discovered in an accelerator. Unfortunately, the mass is so huge that it should make the string discoverable by macroscopic means. I might suggest a sieve. Seriously though, the theorists admit that "all but a few of the vibrational patterns will correspond to extremely heavy particles," meaning particles many times heavier than a grain of dust.[4] It is hard to believe that masses of this sort floating around are undetectable. Greene says that they are unlikely because "such super-heavy particles are usually unstable."[5] It is interesting to note that string theory never says why all such super-heavy particles should be unstable. In fact, there is no theoretical reason they all should be. It is another postulate: postulate #76,904—super-heavy particles are all unstable because if they weren't we might be able to find one. The instability is another axiomatic convenience of the theory.

But let us reverse for just a moment. The tension is even harder to believe than the mass. Try imagining putting a thousand trillion trillion trillion tons of tension on a grain of dust. Talk about tensile strength. Talk about potential energy. And you thought the atom had a lot of stored up energy. What a bomb you could make out of one string! Get one little string to relax for a moment, and you could blow up the entire galaxy. I suspect someone may have miscalculated by a teeny bit. You will say, "C'mon, a whole galaxy? Isn't that hyperbole?" No, 10^{39} tons is actually more than the weight of 2 trillion Suns, which would be *four* Milky Way galaxies. All that tension on one string.

Here's another inelegance. In a sub-chapter ironically entitled, "The More Precise Answer,"[6] Greene develops this idea: the "violent quantum jitters" can be quieted by imagining a collision of point particles as a collision of strings instead. One string represents an electron, say, and the other a positron.

The two strings join for a moment as a string that represents a photon and then re-separate as two new strings. The reason this is an improvement, we are told, has to do with Relativity. Greene uses his two observers George and Gracie (the Burns and Allen ghosts are due massive apologies for being brought into this mess, I think) to "slice" his strings into different events. George sees the strings meeting at one time and Gracie see the meeting at another time. Among all possible observers the meeting point will be smeared out over some time. This smearing calms the quantum jitters.

This is among the most dishonest uses of Relativity and diagramming I have ever seen. In order for his argument to work, Greene has to diagram the strings as three-dimensional objects. For it is not the length of the strings that causes the Relativistic difference in his argument, it is the thickness. But he began the subchapter by admitting that the strings were one-dimensional. He brags that one-dimensional strings can do what zero-dimensional points cannot. Remember that strings have only a length dimension. They have no thickness. As a matter of width or thickness or radius, they act just like points. They have zero radial dimension. This means that Greene's Relativistic slicing is flat wrong. His diagrams are a big fat lie, since they cause you to visualize something that cannot be happening. His words are saying one thing and his diagrams are saying the opposite. If the strings are one-dimensional lines, then the fork where they meet will also be one-dimensional. If you slice it at a dt, then the fork will be in the same exact place for all viewers. String theory adds absolutely nothing to QED or the point problem. It simply adds another layer of lies to cover it up.

Part II

The Math

As promised, I will now critique the math of string theory. String theory has, in the last six or seven years, graduated into M-theory, an 11-dimensional math that attempts to join together the six major 10-dimensional string theories. M-theory has 10 space dimensions and one time dimension, we are told. It is in this matter of dimensions that I have a bombshell to drop. All the extra-dimensional theories that have been proposed since the time of Kaluza in 1919 have contained a basic miSunderstanding of the dimensions they described. No one has seen this before now, but the added dimensions, whether they are Kaluza's one extra dimension or M-theory's seven extra dimensions, are all time dimensions.

To understand why this must be the case, we must go back to the basic calculus. All the higher maths that are used by string theory are based on the calculus. Calculus itself is a math that is based on comparing rates of change. My long paper on the calculus makes this crystal clear, but it has always been understood in some form or another. Velocity is a rate of change of distance and acceleration is a rate of change of velocity. That is why velocity is the first derivative of distance and acceleration is the second. I showed that you can also find third and fourth derivatives of distance, and so on. The third derivative is a change in acceleration and the fourth derivative is a change in that change. These multiple accelerations can really happen: they are physical. I also showed that you could write a velocity in one of two ways, either of which was mathematically acceptable: $\Delta x/\Delta t$ or $\Delta\Delta x$. Likewise, acceleration can be written as $\Delta x/\Delta t^2$ or $\Delta\Delta\Delta x$. A second-degree acceleration can be written $\Delta x/\Delta t^3$ or $\Delta\Delta\Delta\Delta x$, and so on.

You can see that even here there is some sort of mathematical equivalence between x and t. Einstein showed us that this equivalence goes far beyond anything Newton could have imagined, but even in Newton's calculus equations, there was a hidden equivalence. The variables Δx and Δt are the inverse of each other, in some sense. In the equations above, the x goes in the numerator when the t goes in the denominator. This is because, as variables, they always change in inverse proportion, even when no transformational changes are involved. Remember that Einstein showed that as time dilated, length contracted. One gets bigger as the other gets smaller. This is clear in the transform equations of Special Relativity, and it is clear in the equations above as well. A Δx in the numerator is, in some very important sense, the same as a Δt in the denominator.

Why is this important? It is important because what all the big maths of Maxwell's equations, General Relativity, Quantum Mechanics, Kaluza's five-dimensional theory, string theory, and M-theory all do is express fields. All these fields are force fields and force fields are based on some acceleration. By the old equation $F = ma$, if you have a force you have an acceleration. The reason that Kaluza's fifth dimension helped so much at first is that it allowed the expression of both the gravitational field and the electromagnetic field, the only two of the major four that were known at the time. Using the vector fields as they have been defined since the end of the 19th century, the four-vector field could contain only one acceleration. If you tried to express two acceleration fields simultaneously, you got too many (often implicit) time variables showing up in denominators and the equations started imploding. The calculus, as it has been used historically, couldn't flatten out all the accelerations fast enough for the math to make sense of them. What Kaluza did is push the time variable out of the denominator and switch it into another x variable in the numerator, just as I did above. Minowski's new math allowed him to do this without anyone seeing what was really going on.

String theory and M-theory continue to pursue this method. They have two new fields to express, so they have (at least) two new time variables to transport into the numerators of their math. Every time you insert a new variable, you insert a new field. Since they insert the field in the numerator as another x-variable, they assume that it is another space field. But it isn't. It is a transported time variable.

Readers will no doubt be reeling at this information. It was difficult enough to imagine extra space dimensions, most of them curled up like little pillbugs. But how do we make sense out of eight simultaneous time dimensions? It is actually a lot easier than you think, since, once understood, it is easy to visualize. It doesn't take any leaps of faith or warnings that "you can't possibly diagram this, but you must accept it anyway." To show this, I will use the visualization I used in my calculus paper. Let us say you are at the airport, walking along normally. In addition, let us say that you are walking in a perfectly straight line and that your stride is perfect. Each step is the same as every other step. You therefore have a constant velocity, and your stride is, in a sense, measuring off the ground. If you are very retentive, you might even be counting as you walk: 1, 2, 3, 4, and so on. Well then, let us give you a watch, too. Some Swiss-quartz stunner than never misses a beat. So time, the retentive so-and-so that he is, is also counting off his numbers: 1, 2, 3, 4 and so on. Next you come to a moving sidewalk. You step on and for one interval you are accelerated. It is not an instantaneous interval, since some amount of time passes between your initial speed and your final speed. But there is an acceleration over only one interval. It stops at the end of that interval and you have a new constant velocity, a velocity found by adding your own velocity and the velocity of the sidewalk.

Since there was an acceleration over that interval, then by the standard way of expressing acceleration, we how have a t^2 in the denominator: $a = \Delta x/\Delta t^2$. As we know, that can also be written $a = \Delta x/\Delta t/\Delta t$. Either way we have a second time variable. Therefore we might say that we have a second time field and a second time dimension. Now, we must study that interval. What happened over that interval? Did you step into another dimension? Did another dimension open up under you? In a very limited sense, yes. That interval is a sort of sub-interval to the ones you were measuring off with your feet and your watch. But it is not mysterious in any way. It is not curled up anywhere. In fact, you can measure both time dimensions with your watch. That is why we usually just square the time dimension. It is a second measurement over the same interval. If the two time dimensions weren't directly related, we obviously couldn't square them. We would have to call them Δt_1 and Δt_2 or something, and keep them separate.

What I mean when I say that we are measuring two things simultaneously is that we are measuring how far the sidewalk goes in Δt and how far the walker goes in Δt. An acceleration is these two velocities measured over the same interval. So you can see that what we really have is two Δx's and one Δt. But since, in the real world of airports and things like that, we measure strides and lengths on sidewalks with the same measuring rods, it is easier to write the equation with one Δx and two Δt's. Therefore, we get the equation $a = \Delta x/\Delta t^2$.

All this is very elementary, of course. But everyone seems to have lost sight of it at this late date in history. Because what it means, especially when you have a math that is expressed by a lot of superscripted dx's, is that those dx's are not mysterious extra space dimensions, they are equivalent to velocities being measured simultaneously to achieve complex accelerations. These complex accelerations express the meeting of multiple fields over the same (perhaps infinitesimal) interval.

To see this even more clearly, let us say that our man on the moving sidewalk hits yet another field. When he first stepped onto the sidewalk, we might say that he passed through a one-interval field. Well, if we are mischievous, we can plaster that interval with so many fields it will make the man's head spin. We can move him sideways, up, down, or we can spin him any number of revolutions we like. And we can do them all at once. Again, not all at an instant, but all over the same finite interval. Every time we add a motion, whether is a linear motion or a spin, we have hit him with another force. We have also hit him with another variable. We have also hit him with another field. We have also hit him with another dimension. We have done a finite number of things to him during a finite time. Therefore we have created a one-interval field of multiple forces and dimensions.

Stated in this way, there is nothing mysterious about it at all. When we add a dimension, all we have to do is add another Δx to the numerator or another Δt to the denominator (but not both). But none of these extra dimensions is strange or difficult to imagine. We have just imagined it, physically. We could also draw it and diagram it. All we need is tracing paper and overlays. Nothing is curled up. Nothing is unexpressed, nothing only comes out at night.

An 11-dimensional math can be expressed using all the ridiculous equations of M-theory, where super-computers are wasted just storing the postulates. Or you can express it like this: $A = \Delta x, \Delta y, \Delta z / \Delta t^8$.

A perceptive reader may say, "Wait, didn't you just say that we could express a new field or force by either a new space variable in the numerator or a new time variable in the denominator? If this is true, then why can't M-theory call all the new dimensions space dimensions if it wants to?" Well, it can, but it has to be very careful what this implies. I have shown how multiple time dimensions are really a rather simple idea, with no mystery involved. Likewise, seven new space dimensions, correctly interpreted, are also not mysterious or esoteric or difficult to understand at all. A new dimension is a new force applied over a finite interval. If this force is continuous, then it causes a continuous field. By continuous field I mean a field that spreads across an extended set of intervals, not just one interval. Over one interval, a force causes a velocity. Over an extended set of intervals, a continuous force causes a continuous acceleration. We have known this for a long time. But what all this means is that the new dimension is not a new direction in space. Every time we add a new dimension to our math, it does not mean that a new, autonomous x-variable has been invented, going off in some strange new path, like the path of *i* or the path of a curled up pillbug. It just means that we have a new velocity happening simultaneously to all our other velocities over the interval in question. This velocity does not have to have a direction in space all to itself. It does not have to be at a right angle to all previous dimensions. There is nothing to say that forces cannot overlap, or that directions cannot overlap or that times and sub-times cannot superimpose. In fact, they must superimpose (mathematically, they *integrate*). When we create these large-dimensional maths, we are doing so precisely to ask how the various accelerations and forces superimpose. That is what we are seeking. We are seeking the total field at dt, and that total field is a superposition or integration of all the velocities caused by all the force fields present at that interval.

There are two basic and separate ways to "unify" the four known forces and all the spins. One is to find a math that contains them all. The other is to show how one force field is equivalent to another field, thereby simplifying our equations. If we can show that the same basic motion causes two separate fields, we will have unified those two fields. String theory attempts to unify in both ways, but does neither. It tries to unify all fields by expressing them as various vibrations of an ultimate particle, but this part of the theory is just gibberish, as I have shown. It also tries to contain all the fields using a multi-dimensional math, and in this it has made one tiny step in the right direction. If the dimensions are interpreted in the way I have interpreted them above, then they start to make some sense. The first step toward the mathematical expression of a unified total field is to add up all the accelerations and to express them as separate dimensions. But I think it is clear from my airport sidewalk example that a completely successful math must, in the end, recognize the ultimate equivalence of all the time variables. The man in the airport could measure all the various velocities with the same watch, and so can the scientist computing the unified field. When physics understands how all these fields integrate, it will be able to simplify its equations back down to x. y, z, t. It can do this because all the t's are equivalent. Once again, the total physical field, in the presence of eight degrees of linear and angular acceleration, would be $A = \Delta x, \Delta y, \Delta z/\Delta t^8$. That is (a maximum of) 8 fields, but only 4 dimensions.

Some will complain that all the time variables can't be equivalent due to Relativity. But I refer them to my airport example once more. In that example we are studying the field over one interval. In all the maths that are based on the calculus, including tensor calculus and the math of M-theory, we would make that interval an infinitesimal interval or dt. Relativity can't find any variance at dt or dx, for the very simple reason that the observer cannot be any distance away from the phenomena at dx, dt. Notice that in the airport example we have the walker measuring himself using his own watch. He is therefore at no distance from the event and the speed of light has nothing to do with it. The time variables must be equivalent, both physically and definitionally.

A reader will have one final question: why Δt^8? Might that be telling us something fundamental about the number of real accelerations that exist in the universe? Meaning, if five-dimensional math was used for gravity and electromagnetism, then shouldn't the (limited) success of 11-dimensional math be telling us we have 8 fundamental accelerations going on simultaneously, and therefore 8 fundamental force fields?

Maybe. It is possible that we can get to 11 dimensions by adding spin as a dimension wherever we find it. The spin of the electron may be caused by one separate force and the spin of the photon may be caused by another, and so on. Or, some of the accelerations we already know about may be second or third-degree accelerations. No one has ever considered the possibility that the strong force or E/M may be $\Delta x/\Delta t^3$ instead of $\Delta x/\Delta t^2$.

But before we run off pellmell in search of some giant equations to express this, I think we should back up a bit and reassess the entire road to how we got here. In this paper I have shown that string theory is criminally confused about almost everything. Years have been wasted chasing curled up dimensions that don't even exist. There are no Calabi-Yau shapes clinging to the corners of x, y, z. The orbi-folding and all the rest was just mental masturbation. The string theorists have invented shapes and folds and histories and ancestries for these pillbugs nesting in the crannies, and now they find they are completely uninfested. Like some nefarious chemical company, they have soaked the foundations of the communal house in order to roust out the bugs, and now we find that we are all poisoned.

I think it is time to declare that string theory is worthy of a Superfund site and move into a new house. In this house our first order of business is in truly understanding the physical heritage that has come down to us. I have shown in my various papers that there is plenty of work to do in this regard. All the misguided theorists of the past century have quite simply been wrong when they stated, with maximum hubris, that classical and quantum physics was over. Neither classical nor quantum physics is anywhere near finished. We have only touched the surface, even regarding linear maths and "poolball" mechanics. Post quantum theories, whatever they are, will never be possible until QED is corrected and filled out. And QED will never be corrected and filled out until some of the elementary concepts I have spent such a vast amount of time exploring are better understood. Until we understand how our maths are working we can never hope to understand how the universe is working. And this is only the beginning. Velocity, acceleration, circular motion, rate of change, and many other basic physical concepts are not understood to this day. All our physical "knowledge" is dominated by heuristics. Whether we are studying quantum interaction, orbits, or cosmological origins, our equations are overwhelmed by nescience. The best thing to do in this situation is admit the fact and get to work.

A primary piece of this work will be in re-establishing QED without the point particle. One of the only places I agree with string theory is in its critique of the point particle. Quantum math was never able to express its field using an extended particle. String theory realized the problem here and the need to correct it, but it only corrected it by burying it below the Planck limit where no one can see it. In this way the point is given extension, but the extension is only another postulate. Postulate #5 or so: the loop has extension but it is so small that 1) it can't be detected physically, 2) it can't be detected mathematically. Therefore we can fudge over it by misusing the calculus for the millionth time. To my mind this is not a great advance over QED. The only solution is to return to the beginning of QED and start over. We have a lot of very useful heuristics that we can use to guide us, and lots of experimental data. But the math needs a thorough cleaning. The place to begin is in a better understanding of the calculus. Establishing the calculus on the constant differential instead of the diminishing differential will change every physical and mathematical concept of the last 300 years, and will impact all our theories of motion, force, and action. Only once we have rebuilt the old theories from the ground up can we begin counting the force fields and dimensions we will need for a unified math. We may find that we need 11 dimensions. But we may not. We may find that the house looks very different after we have cleared away all the garbage.

Part III

Conclusion

I will end by analyzing a short quote by David Gross, which I also steal from Greene's book. "It used to be that as we were climbing the mountain of nature the experimentalists would lead the way. We lazy theorists would lag behind. . . . We all long for the return of those days. But we theorists might have to take the lead. This is a much more lonely enterprise."[7]

You can almost hear the violins. Those poor put-upon theorists, saving us from the past, leading us bravely into the future. I am not an experimentalist, but when I read this quote my eyes rolled so far back in my head I nearly broke into St. Vitus' dance. The dishonesty literally pours off the page. The string theorist pretending to be an unwilling leader, a humble servant. When in fact he is little more than a shallow revolutionary, a completely monomaniacal, delusional person who has convinced himself that by hoodwinking us he has done us some great favor. Salesmanship posing as magnanimity.

I think you can tell by the tone of this paper that I am angry at string theory, and I don't deny it. The last century would try any honest person's patience, in any number of fields. In my opinion we are past the point of a mild rebuke. The physics department needs a good kick in the pants, and the math department too. Both have degenerated nearly past the point of recognition, and they might as well join up with the art department and begin putting on Dali-esque plays and masked balls. I had hoped that QED would someday develop some humility and that we, as physicists, would get back to work. That we would recognize the huge gaps in our theories, going all the way back to Euclid, and make some effort to fill them. Instead young physicists have continued to learn all the wrong lessons from the recent past and to fail to learn the most-needed lessons. What they have taken from QED is only its Berkeleyan idealism and its intellectual dishonesty. They have remained buried so far under their esoteric maths that they cannot see daylight. And they have continued to dig. They are now at a depth that apparently precludes all cries of logic, all ropes of humility, all ladders of embarrassment.

It seems likely that they will continue to dig until the air runs out. Or until they hit the baby black hole at the center of the Earth, and the self-created chasm at the center of their own theory sucks them into a well-earned hell.

[1] *The Elegant Universe*, p. 123.
[2] Ibid, p. 111.
[3] Ibid, p. 311.
[4] Ibid, p. 151.
[5] Ibid, p. 152.
[6] Ibid, p. 158.
[7] *The Elegant Universe*, p. 214.

A SHORT FORMAL PROOF of GOLDBACH'S CONJECTURE

In a slightly longer paper from 2007, I proved Goldbach's Conjecture with densities and with a visualization. Since densities are just fractions, I now offer this more formal proof for Goldbach stated with simple fractions.

Prove: Any even number can be written as the sum of two prime numbers.
Given: For any even number x , there are x/2 sums, x terms, and x – 1 numbers in the sums (the last sum is always a repeating number).

Depending on the given x, our number of sums may be even or odd, so I will develop separate proofs for each case. If our number of sums is even, x/4 of our sums will be odd-odd sums.

In the sums, it is not terms that meet each other, it is numbers that meet each other. Therefore we must express numbers of primes and non-primes relative to numbers, not to terms. This is important.

Let y = the number of primes less than x.
Let a = the number of primes less than x, as a fraction of *numbers* less than x.
$a = y/(x - 1)$
Let b = the number of non-prime odds less than x, as a fraction of *numbers* less than x.
$b = [(x/2) - y] /(x - 1)$

If Goldbach's Conjecture is false, then no prime will meet a prime. This means that each prime will meet a non-prime odd (NPO) in the sums. The remaining NPO's must meet each other, in which case our number of NPO-NPO sums as a fraction of odd sums would be represented by

$[(x/4) - y]/(x/4)$
or $1 - (4y/x)$.

Notice that if we have no P-P sums, our NPO-NPO sums must be at a minimum. The greatest number possible of NPO's will be covering P's, you see (it helps to look at some real charts to get a feel for this). Since a and b are constant for each x, if we create another NPO-NPO sum, we must create a P-P sum. To create another NPO-NPO sum, we must *uncover* a P.

Now, our fraction of NPO's is in fact $[(x/2) - y] /(x - 1)$, as I have shown. If we subtract away the fraction of primes (the ones covered by NPO's), like this

$$\{[(x/2) - y] /(x - 1)\} - [y/(x - 1)] = [(x/2) - 2y] /(x - 1)$$

Then multiply by 2 to indicate the meeting of two numbers in a sum, like this

$2[(x/2) - 2y] /(x - 1) = (x - 4y) /(x - 1) = [1 - (4y/x)] /[1 - (1/x)]$

We find that our fraction of NPO-NPO meetings should be

$[1 - (4y/x)] /[1 - (1/x)]$, instead of $1 - (4y/x)$

But, $[1 - (4y/x)] /[1 - (1/x)] > 1 - (4y/x)$

Therefore, the term $1 - (4y/x)$ is disallowed. It is disallowed for this reason:

Again, the creation of a P-P sum will create *more* NPO-NPO sums; so when P-P is forbidden, the fraction of NPO-NPO sums will be at a minimum.

Therefore, the minimum fraction of NPO-NPO sums for any given values of x and y is $[1 - (4y/x)] /[1 - (1/x)]$. $1 - (4y/x)$ is less than the minimum, therefore it is mathematically disallowed.

Because the number of sums must be an integer value, the minimum fraction of NPO-NPO sums for any given value of x and y cannot be $[(x/4) - y]/(x/4)$. In order to satisfy our actual minimum, it must be $[(x/4) - y + 1]/(x/4)$. Notice the plus 1. We must add an NPO-NPO sum to satisfy our calculated minimum.

Since a and b are constant, and cannot change for a given x, you cannot create another NPO-NPO sum without also creating a P-P sum. We have just been forced to create one P-P sum.

This proof works for all values of y and x, no matter how small the fraction y/x becomes. It works for all prime densities.

This proves Goldbach's Conjecture for even numbers with an even number of sums.

This simple solution is made possible by the fact that the last sum is always composed of a repeating number. As you now see, this means we have one fewer numbers in our charts than we have terms, skewing the fractions up. Because our number fractions are slightly higher than our term fractions, our minimum meetings are affected. The simple math shows that to meet this new minimum, we must add 1 to the term minimum. And if we do that, we must create a prime pair. If you get lost in the math, remember that it is all caused by the inequality between numbers and terms. That is, it is caused by the fact that the last sum is always a repeater. If our given even number is 62, for example, the last sum is 31+31. We have more terms below 62 than we have numbers.

Now we will look at even numbers with an odd number of sums.

If our number of sums is odd, $(x + 2)/4$ of our sums will be odd-odd sums.

$a = y/(x - 1)$
$b = \{(x + 2)/2] - y\} /(x - 1)$

If Goldbach's Conjecture is false, then no prime will meet a prime. This means that each prime will meet a non-prime odd (NPO) in the sums, in which case our number of NPO-NPO sums as a fraction of odd sums is represented by

$\{[(x + 2)/4] - y\}/[(x + 2)/4]$
or $(x + 2 - 4y)/(x + 2)$

But our fraction of NPO's is $\{(x + 2)/2] - y\}/(x - 1)$. If we subtract away the fraction of primes (the ones covered by NPO's), like this
$\{[(x + 2)/2] - y\} /(x - 1)\} - [y/(x - 1)]=\{[(x + 2)/2] - 2y\} /(x - 1)$

Then multiply by 2 to indicate the meeting of two numbers in a sum, like this

$$2\{[(x + 2)/2] - 2y\} /(x - 1) = (x + 2 - 4y) /(x - 1)$$

We find that

$$(x + 2 - 4y) /(x - 1) > (x + 2 - 4y)/(x + 2)$$

Therefore, the term $(x + 2 - 4y)/(x + 2)$ is disallowed.

This proves Goldbach's Conjecture for even numbers with an odd number of sums.

ABOUT THE AUTHOR

You may read about the author at this website:
http://mileswmathis.com/bio.html

To read other papers on physics and mathematics, please go to
http://milesmathis.com

At publication time, there were over 1,500 pages of papers at this site.

Please note those are two different websites.

Made in the USA
Columbia, SC
22 January 2022

54632996R00219